Contemporary Meanings in Physical Geography

From what to why?

This book is dedicated to the memory of
Dick Chorley, 1927–2002,
and also to what he recognized as really important:
the next and future generations of physical geographers
and their new ideas.

Contemporary Meanings in Physical Geography

From what to why?

Edited by

Stephen Trudgill

Department of Geography, University of Cambridge (UK)

and

André Roy

Département de Géographie, Université de Montréal (Canada)

A member of the Hodder Headline Group
LONDON
Distributed in the United States of America by
Oxford University Press Inc., New York

First published in Great Britain in 2003 by
Arnold, a member of the Hodder Headline Group,
338 Euston Road, London NW1 3BH

http://www.arnoldpublishers.com

Distributed in the United States of America by
Oxford University Press Inc.
198 Madison Avenue, New York, NY10016

British Library Cataloguing in Publication Data
A catalogue record for this book is available from the British Library

Library of Congress Cataloging-in-Publication Data
A catalog record for this book is available from the Library of Congress

ISBN 0 340 80689 3 (hb)
ISBN 0 340 80690 7 (pb)

1 2 3 4 5 6 7 8 9 10

Typeset in 10/14pt Gill Light by Phoenix Photosetting, Chatham, Kent
Printed and bound in Great Britian by MPG Books Ltd, Bodmin, Cornwall

What do you think about this book? Or any other Arnold title?
Please send your comments to feedback.arnold@hodder.co.uk

Contents

Editor's preface

This is a book that explores how the discourses of the personal meanings of researchers, their scientific endeavour, the academic contexts of the day and cultural significances are interwoven to provide the contemporary meanings of physical geography.

Scientists in their work are motivated by a number of personal meanings and specific scientific contexts and they are also in a relationship with the sociologies of the day. At a time when physical geography has a place in schools, universities and research institutions which is less and less secure, and when science itself is becoming widely scrutinized, it is apposite to examine these personal motivations, contexts and sociologies. It seems from the authors' contributions that physical geography is changing its meanings, but is alive and kicking.

This text is written by experienced physical geographers for undergraduates who may be thinking of embarking on their own enquiries or indeed may be considering whether to actually study physical geography or not. It will also be of interest to postgraduates and others undertaking further physical geography research. This is then a rather different kind of physical geography book, comprising a collection of international writers' thoughts which reveal personal motivations, the excitement and interest, and looks at tensions in the worlds of meaning in which physical geography is involved. How are meanings derived from an understanding of the physical environment related to meanings which derive from the contexts of applications and environmental issues? Is the future of physical geography one where the only, or at least the dominant, meanings are framed in the contexts of environmental issues?

The prime motivation for any research worker is a sense of enquiry together with a curiosity about the world around them. But how do we come to frame our questions and how far do our questions influence our answers? Involved is a sense of logic or order, whereby we make sense of the world – interest and meaning often deriving from a sense of connectivity and relationship. However, does each sense of order we derive then constrain our next enquiry and deny a richness through abstraction? Is there a tyranny of models or even a paradigm of paradigms? Have we moved from an era of interpretation through a time of measurement and quantification to enter an era where, on the one hand, only chaos and complexity are seen in the physical environment while, on the other, only the contexts of social relevance and environmental issues are seen as providing legitimate meanings? What is the wider cultural significance of physical geography?

In SETTING THE SCENE, the authors lay out the history of the development of physical geography and assess some of the challenges inherent in looking at meanings. In PERSONAL MEANINGS, authors then write autobiographically about how they became interested in physical geography and what it means to them, and reflect on what they have learnt by looking back. The core of the book is an examination of RESEARCH MEANINGS, with specific considerations of concepts and applications in research. Finally, while most contributors have already concluded their chapters with a sense of advocacy or challenge, in FUTURES, two authors assess some specific challenges and directions concerned with the importance of communication and the interaction of personal, scientific, contextual and cultural meanings.

Authors

Bayliss-Smith, Tim. Department of Geography, University of Cambridge, CB2 3EN, UK.

Bobbo, Gerardo. UNAM, Rey Inchatiro 355, Colonia Vista Bella, 58090, Morelia, Mexico.

Burt, Tim. Department of Geography, University of Durham, DH1 3LE, UK.

de Boar, Dirk. Department of Geography, University of Saskatchwean, 9 Campus Drive, Saskatoon S7N 5A5, Canada.

Favis-Mortlock, David. School of Geography, Queen's University, Belfast T7 1NN, Northern Ireland, UK.

Keylock, Chris. School of Geography, University of Leeds, LS2 9JT, UK.

Lane, Stuart. School of Geography, University of Leeds, LS2 9JT, UK.

McGregor, Glenn. School of Geography, Earth and Environmental Sciences, University of Birmingham, B15 2TT, UK.

Pulido, Juan. CRUCO, Mexico.

Rhoads, Bruce. Department of Geography, University of Illinois, Urbana, IL 61801, USA.

Richards, Keith. Department of Geography, University of Cambridge, CB2 3EN, UK.

Roy, André. Département de Géographie, Université de Montréal, C.P. Centre-Ville, Montréal, Quebéc, H3C 3J7, Canada.

Sims, Peter. Department of Geography, School of Geography and Genology, University of Plymouth, PL4 8AA, UK.

Thornes, John. School of Geography, Earth and Environmental Sciences, University of Birmingham, B15 2TT, UK.

Trudgill, Stephen. Department of Geography, University of Cambridge, CB2 3EN, UK.

Urban, Mike. Department of Geography, University of Missouri, Columba, MO, USA.

Viles, Heather. School of Geography and the Environment, University of Oxford, OX1 3TB, UK.

Acknowledgements

This book began as a serviette. André Roy and I were having lunch with Alistair Kirkbride in the Cambridge Graduate Centre when I began to talk about the possibilities of a new book on physical geography which the publishers had asked me to write. André rapidly proposed that it was the personal meanings of physical geography which were important and I hastily began scribbling 'personal meanings' on the nearest available piece of paper, and thus he became the co-editor, with much of his basic idea imbuing the book.

The editors would like to thank Alistair for his insights at the original lunch which provided the impetus for the nature of the book, the authors for their contributions and colleagues at Cambridge and elsewhere for reading the chapters and making comments on the drafts. We thank Ron Cooke for providing the preface. Also the publishers for their patience – as usual, I have sworn that I will never edit another book; such are the pressures on contributors that deadlines have been set and re-set and re-set more times than I care to remember, and this not without regret about those authors who had signed up to write but never made the final, final, final deadline.

AGR would like to thank Keith Richards and Emmanuel College for the opportunity to spend time in Cambridge in 1999. It is during this visit that ideas for the book were developed.

STT would like to thank, as ever, students in supervisions and colleagues in the Cambridge Geography Department and in Robinson College who sustain an all-important ethos of intellectual debate. Dick Chorley was to have written a postscript for this book as his personal assessment of recent changes in physical geography, but he died before he could do so: we shall never know what a blend of insight, humorous anecdotes, wisdom and scurrilous wit that assessment might have been – though I am pleased that it was Dick who characteristically found me the 'geomorphological poems' on glaciers by Benson and Ruskin which appear in the text.

PREFACE

Very few scientists indulge in autobiography, and even fewer are sufficiently distinguished to command the attention of biographers. Alas, the motivation, vision, enthusiasm and inspiration of most scientists is not a matter of public record. That is a great pity, because the insights we can gain from the personal experience of our colleagues can be invaluable as a way of understanding both ourselves and the scientific questions we ask. Simon Winchester's (2001) account of William Smith, for example, revealed how this driven, brilliant, self-trained scientist had overcome monumental obstacles across the path leading to a geological map of England and Wales, obstacles that would have stopped most of us: bankruptcy, imprisonment, plagiarism and the not uncommon disdain of his contemporaries, not to mention the intellectual difficulties of interpreting for the first time the order of the rocks.

Physical geographers, as a species of the genus field scientist, share with all scientists a natural curiosity and an interest in formulating answerable questions. But the questions, and the initial inspiration needed to prompt them, are often distinctive. We are concerned both with explaining our physical environment and understanding how human communities interact with it. That conjunction is more important and difficult to study than it might seem: explaining the physical environment prompts one set of questions; understanding the links between the environment and communities another. We share both questions with others. For example, we can ask with Bob Dylan 'How many years can a mountain exist, before it is washed to the sea?'; and in addition to contemplating the evolution of mountains, we can also recognize, along with William Wordsworth and others at the end of the 18th century, that to many people mountains as a resource can prompt reactions of 'gloom' and 'glory' and much more besides.

Physical geographers, it seems to me, often gain their initial inspiration from the landscape and the processes related to it. For us, process is generic – water really does usually, but not always, flow downhill, and the pebble in a stream is battered by predictable forces. But location is unique, and specific landscapes always require interpretation that depends in part on the unique consequences of unique historical events: it is for this reason that what is observed so often conflicts with what is known, and new questions that inspire research come flooding in. We all seek that brutal confrontation, and we hope it happens often.

Unfortunately, many choose not to record experiences that lead to the biography of disciplines: they usually avoid such monstrous concepts as paradigms and paradigm shifts in favour of ploughing on and extending existing knowledge.

I first remember it in the Atacama Desert, when my colleagues left me alone on a broad, apparently featureless desert plain, far from any settlement. (I believed, of course, that they would return!) There was nothing for it but to examine the stony surface – a desert pavement. After a while, it became clear that this surface accumulation of stones could not have been formed by wind removing the fine materials, as all the textbooks said, for a variety of reasons. So if not deflation, what were the processes of stone concentration at work? That question seemed very exciting to an undergraduate; and so indeed it was, because it led to research that ultimately convinced me that stones on desert surfaces are a crucial archive of surface dynamics, and that they are invaluable to understanding ground conditions that, in turn, are vitally important for sensible human development.

If, for physical geographers, process is generic and location is unique, scale is fundamental, because we very often move from the inspiration provided by a specific location or process to an understanding of their relationships that ultimately encompass the Earth as a whole. So, at my desert pavement, there is movement of dust. That local movement is part of a regional pattern of aeolian transport which, in turn, contributes to global patterns of sediment yield.

My desert patch in Chile has a personal meaning, and it contributes to a research meaning. The site reminds me of another autobiographical context that is rarely public, but which brings the two meanings together. For many of us who work in remote and dangerous environments, a sense of survival co-exists with the search for scientific understanding, but the act of exploration no longer attracts the same attention as the process of scientific discovery. That's a pity too. For example, in the writings of Ralph Bagnold, in the *Geographical* journal and elsewhere, we see a brilliant geomorphologist creating a stunning understanding of desert dunes, developing the means of desert survival and, later, means of waging desert warfare.

So this book combines autobiography of physical geographers with biographies of physical geography. Both are fascinating and help us to see why our physical environment is such an exciting challenge to us, as scientists, to students beginning their studies, and to anyone dedicated to understanding the Earth as the human habitat.

Professor Sir Ron Cooke
Vice-Chancellor, University of York
President Royal Geographical Society
October 2001

PART I

SETTING THE SCENE

from Trudgill and Roy, ed.
Contemporary Meanings in Physical Geography
(Arnold, 2003)

1
Previous actors and current influences:
trends and fashions in physical geography

Peter Sims

1.1 Introduction

At first sight, this chapter is a straightforward introduction to the nature and content of the major themes and issues that have formed the basis of modern physical geography from the end of the 19th century to the present day. However, a little further thought poses two key questions. First, exactly what should be included in the term physical geography? Second, should the approach be essentially chronological or concentrate on thematic approaches and specific sub-disciplines? Being conscious of a recent comprehensive review of *Physical Geography* (Gregory, 2000), it would be impossible in one small chapter to do full justice to all branches of the subject – itself so heavily dominated by geomorphology and its own numerous subdivisions. Nevertheless, I have attempted to include some aspects of biogeography, climatology, soils, hydrology and Quaternary studies. However, most consideration is given to the links through to geomorphology – a reflection not only of the overriding influence of this area of study, but also of my own training and experience.

What follows is the result of an early decision on my part not to present a strictly chronological discussion, but to merge the historical development of physical geography with an identification of some of the key ideas which have emerged and been developed and modified. I attempt, therefore, to explore the way in which certain ideas have created a chronology of the development of this particular branch of science, and also examine how some ideas have become recurring themes which have been worked and re-worked through time. One could of course refer to these in terms of paradigms and paradigm shifts (or even paradigms re-visited), although I prefer not to label matters in this way. This may perhaps be a result of my own uncertainty about what rightly can be given such a status (or not), as the case may be!

1.2 Speculation, observation, classification and categorization

The foundations of modern physical geography – origins and evolution: the Darwinian inheritance. The cycle of erosion and the influence of W.M. Davis – denudation chronology and all that! Climaxes, succession and stages. Climatic zones and biomes. Experimentation – finer tuning of the Davisian 'normal' concepts.

Davies (1968) chronicled the very early ideas in earth sciences, reviewing the work of various authors who attempted explanations of earth history. However, the true beginnings of modern physical geography can be traced to the numerous developments in scientific thinking that emerged at the end of the 18th century and which were significantly advanced during the 19th century. Of particular note were the publications of a number of geologists, e.g. James Hutton's *Theory of the Earth* (1795).

For Hutton, the earth was a machine – a kind of three-phase engine in which denudation of continental rocks provided the soil to maintain the earth's fertile mantle, where continental debris was transported seawards by rivers in a second phase and sedimentation which took place on the ocean floors. There the material was converted into new sedimentary strata, eventually to be uplifted to form new continental masses in the third and final stage of the process. This pattern could be repeated through time, and so we see the beginnings of a concept which was to sustain much geomorphological thinking through to the middle of the 20th century i.e. repeated cycles of uplift, denudation resulting from the dominance of fluvial erosion and the subsequent deposition of transported rock materials. Hutton maintained quite clearly that the origin of the earth pre-dated existing continents and, in so doing, established long timescales which effectively banished earlier ideas which relied heavily on biblical interpretations of the creation of the earth and the role of a catastrophic universal flood.

Hutton's work did not receive much in the way of immediate acclaim or recognition. Davies (1968) explains that apart from the problems caused by the dominance of accepted Christian teaching on the origin of the earth, Britain was embroiled in concerns surrounding the revolution in France – it would seem that the 'Establishment' could not handle two revolutions at once! Hutton's thesis was re-stated and condensed by Playfair (1802) in his *Illustrations of the Huttonian Theory of the Earth*, but it required the input of further ideas before the old accepted theories were finally brought seriously under threat.

In part, momentum to these changing notions was given by the concept of uniformitarianism expounded by Charles Lyell in his *Principles of Geology* (1830) – 'being an attempt to explain the former changes of the Earth's surface by reference to causes now in action'. Lyell, throughout the following 45 years and 12 editions of his text, continued to extend the hypothesis that the earth's surface was subject to natural processes operating over long timescales and that many of these processes could be observed moulding the scenery of the present day. Thus the present became the 'key' to the past and another fundamental geomorphological axiom was established, although the scale of operation and frequency of events were not fully discussed as possible 'flaws' in the general acceptance of these ideas.

The final impetus to this 'new' thinking came in the latter half of the century with the publication in 1859 of Charles Darwin's *Origin of Species*. The impact on physical geography was immense. First came the recognition that the earth's surface 'evolved', i.e. it changed through time – ideas developed earlier in Darwin's (1842) work on coral islands, and second came a concept of organisation – that there were distinct relationships between plants and animals and the environment.

All of these ideas moved Victorian science forward significantly, although basic observation and description and, in some cases, the somewhat routine collection and categorization of natural phenomena continued for some time, as exemplified in works such as Mantell's *Geological Excursions round the Isle of Wight* (1854) and Dana's *Manual of Mineralogy and Lithology* (1879). Indeed, it was not until the last 15 years or so of the 19th century that the earth's landscapes began to be explained utilizing some of the newer ideas – see for example *Geology: chemical, physical and stratigraphical* (Prestwich, 1886).

It was exactly at this stage that one of the great luminaries of geomorphology enters upon the scene. W.M. Davis (1850–1934), working mainly in the eastern part of the USA, absorbed and developed evolutionary and uniformitarian ideas into his explanation of landforms and the cycle of erosion (and related complications or modifications to the original hypothesis) in a number of papers published from 1889 to 1906 (see *Geographical Essays* edited by D.W. Johnson, 1954).

Davis suggested that after rapid uplift of an emergent sea floor, fluvial erosion (the dominant **process**) would act through time upon the underlying geology (lithology and **structure**) to produce a landscape described in terms of **stage** of development (youth, maturity and old age). An end-point would be a fluvially eroded base-level surface (the peneplain). To this basic model he introduced modifications to allow for rejuvenation (renewed uplift) and further hypotheses on the evolution of the meandering habit of rivers, together with the concept of grade and the graded profile in which the denudation of valley slopes was balanced by the removal of eroded material by rivers. In this context, according to Higgins (1975), Davis synthesized contemporary geological thought in using and developing the ideas of other American earth scientists, namely Powell (1834–1902) and Gilbert (1843–1918). Later work by Davis accepted that not all landscapes could be attributed to the basic model and papers on arid cycles and marine cycles were added, as were the 'accidents', which accounted for volcanic and glacial landforms.

Davis' contribution to geomorphology was truly monumental. His thinking and writing dominated the subject for over 50 years and established geomorphology as the dominant branch of physical geography. In the States, Davis' disciples included D.W. Johnson (who edited the main essays in 1909), Von Engeln (1942) and Thornbury (1954). Outside the US, Davis' ideas were championed (amongst others) by Wooldridge and Morgan (1937) and Wooldridge and Linton (1939), based in Britain, and later by Cotton (1941,1942), working in New Zealand.

Chorley et al. (1973) describe Davis as being 'too important and too prolific to be ignored'. In fact, his publications, stretching over a 58-year time span, averaged about 10 per year and included significant output on a variety of themes in geology, meteorology, oceanography and geographical education, as well as the well-documented series on the genetic evolution of fluvially dominated landforms – the so-called 'normal cycle of erosion'.

In other branches of physical geography similar developments, based partly on Darwinian 'evolutionary' ideas, occurred. In biogeography, Clements (1916, 1928) explained that the distribution of plant species in space and time was the result of a succession as the plant community adapted to sets of environmental or controlling conditions, ultimately producing a 'climax' vegetation community subject to a main control – climate; whilst Bjerknes (part of an influential early 20th century school of Norwegian meteorologists) discussed the life cycle of mid-latitude depressions – introducing the important concepts of cyclogenesis and frontogenesis.

1.3 Theorization and 'love affairs' in geomorphology

Denudation chronology, dynamic equilibrium, slopes and general systems. Climatic geomorphology. The beginnings of quantification and the shift towards processes.

In many respects, the theory of the normal cycle of erosion was unproven. Although it was based on extensive and detailed field observations of landforms in the eastern USA and Europe, it remained essentially a descriptive conceptual model. Davis, of course, was not without his critics, and perhaps the most notable objections to his ideas came from the disbelief that such rapid uplift could occur to initiate a new cycle and that examples of the end-point (the peneplain) could not be identified at or near any current base level of erosion.

These objections were strongly expressed by W. Penck (1924), who observed rectilinear (straight) rather than convexo-concave slopes and who was developing ideas on parallel slope retreat from fieldwork in South America and elsewhere. To Penck, slope form and erosion were controlled primarily by varying rates of uplift and the 'balance' achieved between the rate of down-cutting of a valley floor and the angle of the valley-side slope. Slopes were thus independent of a cycle (and the stage reached within it) but were in a state of dynamic equilibrium controlled by the energy of a river in relation to the rate of uplift of the land surface. Penck's ideas were taken up by Wood (1942) and later embraced by King (1950, 1953) working in South Africa, who presented 50 statements (or Canons) about landscape evolution – the end-point of which was a landform termed a pediplain. According to Ollier (1995), 'King set the scene for an unfortunate period of Davis-bashing, in which theorists ridiculed the old cyclical ideas but did not (like King) provide alternatives'.

Nevertheless, in Britain, Penck's views (originally expressed in German) failed to gain much support – which Brown and Waters (1974) attributed to 'the linguistic ignorance of most British geomorphologists' – and the Davisian model continued to be 'established' teaching well into the 1950s. Such was the nature of the influence that it initiated substantial research, generally grouped under the heading of 'denudation chronology', from numerous workers in which local and regional landscapes were allotted evolutionary histories.

The scale of the output from Britain was prodigious, as demonstrated in the extensive listings in *A Bibliography of British Geomorphology* (Clayton, 1964). In fact, the work of Davis was still demanding treatment well into the 70s – as is seen by the chapters devoted to his cycle in Small's *The Study of Landforms* (1970) and in Sparks' *Geomorphology* (1972). Sparks remarked:

> *The importance of Davis is not whether he was right or wrong. It would be a sad reflection on our subject if, after three-quarters of a century, we found nothing to criticise or disbelieve. But, Davis drew attention to the main features of landform development and suggested an overall scheme to embrace them, so providing the dogma to accommodate the expression of our doubts.*

Such a comment was of course made in the light of a number of further ideas that had again emanated from the USA. In part, a transition had been initiated by Horton's (1945) work on the erosional development of streams and their drainage basins, and Strahler (1950a, b) had advocated a quantitative approach in which landforms would be measured and erosional processes analysed. Thus was founded a 'school' of American fluvial geomorphology as demonstrated by the further work of Strahler (1952), Schumm (1956), Hack (1960), Wolman and Miller (1960), and which culminated in the publication of Leopold, Wolman and Miller's *Fluvial Processes in Geomorphology* (1964) – a text which was the foundation of the shift towards laboratory and field experimentation inherent in the emerging area of process geomorphology. Meanwhile, in Britain, considerable work had been initiated on the mapping of slopes: for example, Savigear (1952), Waters (1958) and the morphometric analysis demonstrated by Chorley (1957) 'heralded in' a new thinking which was to establish geomorphology as a quantitative science.

Other developments were in progress at around this time. Peltier (1950) had specified a number of 'morphogenetic regions' characterized by differing climatic controls (based on mean annual temperature and rainfall) in which frost action, chemical weathering, mass movement, wind action and pluvial erosion all played a part. This was the essential foundation to his discussion of a periglacial cycle of erosion – thus extending the Davisian theme. Problems inherent in Peltier's scheme relate to the 'broad brush' that mean annual data present, with no consideration of soil moisture and rainfall/run-off relationships or the magnitude and frequency of geomorphological events.

Climatic geomorphology also had other protagonists. In France, Tricart and Cailleux (1972) established morphoclimatic zones which incorporated soil and vegetation phenomena, although these relied heavily on the latitudinal classification of regions rather than on detailed data relating to the operation or nature of processes within the delineated zones. In Germany, Budel (1963) had developed climatogenetic geomorphology in which each zone displayed particular landscapes – the result of differing landscape-forming processes.

Stoddart (1969) in his review of climatic geomorphology concluded that: 'As present formulated, much climatic geomorphology is (a) not new, being implicit in Davis's own work; (b) not well established, especially in terms of the climatic control of processes; and (c) premature, in setting up generalised world-wide schemes ...' and in a further critique of the topic, Twidale and Lageat (1994) comment that: 'Climate is certainly not an overriding consideration in the interpretation of landscape.'

Nevertheless, interest in climatic geomorphology did initiate a number of more detailed studies of weathering and erosional processes operating in differing areas, as exemplified by the work of Berry and Ruxton (1959), Corbel (1959a, b), Ruxton and Berry (1957, 1959), Thomas (1965, 1966, 1974), Douglas (1967a, b) and Ollier (1969). Even the great debate on the nature of tors and granite landscapes in Britain (Linton, 1955; Palmer and Neilson, 1962; Brunsden 1964; Waters, 1964; Eden and Green, 1971; Green and Eden, 1971) had much to do with the operation of geomorphological processes under differing climatic regimes through time.

Explanations of slope form and slope processes were also a feature of a shift towards a more thorough interpretation of the role of processes (rather than structure or time) in landscape studies. Strahler (1950a, b) embarked on quantitative description and analysis which generated much data collection and interpretation, especially in Britain (see Brunsden, 1971 and Brown & Waters, 1974). One of the key workers in this arena was A. Young who, in a series of papers between 1960 and 1971, explored relationships interlinking soils, regolith formation, denudation and slope form. The research was incorporated into his text *Slopes*, published in 1972. A further systematic (and somewhat quantitative) approach to the topic was presented by Carson and Kirkby – also in 1972. They examined slope development through time, soil and rock instability, surface and sub-surface erosion and rates of operation and formation as applied to different climatic regimes.

The final 'affair' included in this section is that with the systems approach. In part, the move towards quantification inherent in the empirical work of the American fluvial school and the shift away from descriptive denudation chronology elsewhere represented the attractive 'chemistry' which pulled geomorphology into this new liaison. Systems theory was seen as providing a methodology (Stoddart, 1967) around which all aspects of physical geography could unite and, in so doing, join the 'mainstream of scientific progress'. In some respects, geomorphology was the last branch of physical geography to become attracted to this new thinking – probably because many workers were somewhat reluctant to leave behind their embracement of the Davisian ideal. Chorley (1962) had 'paved the way', but the real shift came with the publication of *Physical Geography: A Systems Approach* by Chorley and Kennedy in 1971.

Thus morphological, cascading, process-response and control systems were all defined and exemplified, which in turn generated a plethora of further research in nearly all aspects of geomorphology, which had by now firmly 'nailed' the process flag to its mast. One further point is of note here: Chorley and Kennedy had intended that the 1971 text would present a view of landscape and processes in terms relevant to the human geographer. I would not presume to comment on the success or otherwise of this aim, although subsequently Bennett and Chorley (1978) furthered the systems approach in providing an interdisciplinary convergence on environmental matters. They integrated socio-economic aspects with physical and biological theory and in so doing spawned work on a range of environmental matters.

1.4 Measurement, empiricism and monitoring

Processes – the 'key' to unlocking the truth about landforms and landscape evolution or the 'tunnel vision' of reductionist science? Hydrology and small catchment experiments, drainage basins – form, pattern and processes ancient and modern.

Process studies became very much 'The Holy Grail' of geomorphology from the late 60s onwards. The work of the American fluvial geomorphologists had set the scene, and in Britain influences were also coming from the fast developing science of hydrology (see, for example, Ward, 1967). Drainage basins were seen as fundamental landscape units (Chorley, 1969a), whilst catchments and catchment studies formed a framework upon which numerous aspects of pure and applied research were founded.

Chorley's edited text, *Water, Earth and Man* (1969b), established the genre of the time. It also addressed many applied aspects of hydrology and fluvial geomorphology and discussed the relationships across 'the divide' between physical and human geography. In some respects this important contribution to the literature of physical geography also 'heralded' the rise of environmental concerns which were to establish themselves from the 1970s onwards. The scope of *Water, Earth and Man*, with its strong underpinning of physical hydrology, presented a model for later works such as Gregory and Walling's *Drainage Basin: Form and Process* (1973) and the important research paper collections: *Fluvial Processes in Instrumented Watersheds* (Gregory and Walling (eds), 1974), *Geographical Approaches to Fluvial Processes* (Pitty (ed.), 1978), *Man's Impact on the Hydrological Cycle in the United Kingdom* (Hollis (ed.), 1979), and *Catchment Experiments in Fluvial Geomorphology* (Burt and Walling (eds.), 1984).

This predominant interest in fluvial processes can be attributed to a number of influences. First was the realization that many geomorphologists had spent, perhaps, too long describing humid temperate landforms, essentially from 'structural' and 'evolutionary' views, whilst largely omitting any attempt to understand the processes involved. Second, that techniques were becoming available to establish short-, medium-

and longer-term field experiments in which processes and detailed changes in the natural environment could be studied at a range of scales, as illustrated in a number of the British Geomorphological Research Group Technical Bulletins published between 1969 and 1979. These 'technical' responses were also partially a reaction to influences coming from outside the academic world, such as the thrust given to environmental monitoring by the 1963 Water Resources Act in England and Wales (with its demand for nation-wide hydrometric networks) and also to the launching by UNESCO in 1965 of the first International Hydrological Decade. Thus fluvial studies on hillslopes, in small catchments and in larger drainage basins, were 'driven' by both pure and applied research initiatives aided by fast developing techniques for environmental monitoring, data handling and storage.

On this latter point, in a discussion of hydrological and fluvial processes, Walling (1987) noted that the impact of technological advances on process studies could either enable more intensive investigations of phenomena (i.e. the reductionist approach where more and more detail is revealed, investigated and understood at larger and larger scales for smaller and smaller areas) or inform in the formulation of new problems and complex inter-relationships between suites of processes, thus seeding new research questions. In some respects, these approaches 'move' in opposite directions in terms of scale of interest, locational inclusivity and the comprehension of which processes control natural environmental systems.

A further element of fluvially based process studies was the renewed interest in channel form and pattern, the solute and sediment loads of rivers and in the controls surrounding denudation of the earth's surface. Examples are provided by the work of Richards (1982), Walling and Webb (1983) and Knighton (1984). New technologies and methodologies were also helping with these approaches in, for example, the use of magnetic imprints to 'link' sediment sources with rates of erosion or sedimentation (Oldfield, 1983) or in the use of Caesium-137 and Lead-210 as 'tracers' (Wise, 1980; Hadley and Walling, 1984). Many of these developments gave insight into rates of operation of the processes effecting landscape change. However, as Goudie (1995) pointed out, some care is required in extrapolating the effects from data derived from short- or even medium-term monitoring – especially where the influence of humans in the landscape is involved.

Despite the apparent dominance of fluvially based studies, research into processes was not limited just to these systems. King's second edition of *Beaches and Coasts* was published in 1972 and demonstrated the importance of detailed studies of physical oceanographic processes (suitably modified for geomorphological research in the nearshore zone) in explaining resultant coastal landforms and sediment movement. Similarly, Pethick's (1984) *Introduction to Coastal Geomorphology* 'attempts to bring coastal geomorphology into the established framework of process studies'. In fact, virtually every identifiable branch of geomorphology was to generate a major text (or texts) specifically related to a delineated area of study in which the role of processes was to be amply highlighted. Of note here are Cooke and Warren's *Geomorphology in Deserts* (1973), Embleton and King's *Glacial Geomorphology* (1975), French's *The Periglacial Environment* (1976), Sugden and John's *Glaciers and Landscape* (1976), Washburn's *Geocryology* (1979) and Derbyshire, Gregory and Hails' *Geomorphological Processes* (1979). To this list should be added the important essay compilation, *Process in Geomorphology,* edited by Embleton and Thornes (1979).

The progress and 'control' that process studies exerted continued, 'increasing the fragmentation of the branches of geomorphology' (Gregory, 2000). There were even accusations that process-orientated geomorphologists had 'lost the plot' and forgotten that the main purpose of process studies was to explain the origins of landforms (Church, 1980; Conacher, 1988). Nevertheless, the trend is ongoing and there remains considerable division between the reductionist (processes in their own right) approach and those who see processes as either an essential part of an overall picture of landscape evolution or as a necessary underpinning to applied studies which attempt to address the role of human influences on natural systems.

1.5 Modelling and the modelling ethic

Armchair theorizing. Hands-on with hardware – monitoring and measuring, creating the data. The numerical explosion – nice work if you can hack it! New explanations – chaos and fractals.

In the sense that a model is a categorized and simplified representation of the real world, models have been with us for a considerable time. Even the Davisian landscape evolution treatise can be seen as a conceptual model which communicated 'an explanatory scheme for teaching which achieved its great success through its relative simplicity and flexibility' (Kirkby, 1987). However, models have been more usually associated with the shift from qualitative description to quantification, which developed from the studies of the functioning of natural systems (led by the American Fluvial School) and the morphometric expression of both drainage networks and the slopes across which the fluvial processes operated. The extensive and detailed studies of geomorphological processes emanating from laboratory and field experiments and the numerical data such research produced were further aided by two factors: first, the significant shift to statistical methods to 'sort' and hopefully give explanation to the data (see, for example, Doornkamp and King's *Numerical Analysis in Geomorphology* (1971)), and second, the application of rapidly expanding computing hardware and software power – the impact of which has been almost beyond measure.

But to return to beginnings, as Board (1967) ably demonstrated, maps were probably the most significant early examples of landscape models that incorporated both elements of direct observation and field measurement (if only a rough estimate of distance and compass bearing between two places or features). Early maps also involved a parameterization of the model boundaries by the selection of a suitable scale and associated cartographic decisions on what to show or leave out of the finished product (model). Viewed in this way, the map had real applicability in that it inherently had clear objectives and was also an effective means of communicating information about the real world – characteristics that all good models should demonstrate. The limitations were obvious however, in that reliance was placed on inadequate data (surveying and measurement) and excessive 'artistic licence' – an early example, perhaps, of the 'all hypothesis and little data' syndrome! Maps however, are more than abstract concepts and were to become an important tool in representing geomorphological features, not only in terms of landscape outline or topography, but also as an interpretative methodology extensively used for detailed morphometric analysis and explanation of landforms – a fashion which seems to have passed its acme.

Modelling has in many ways reflected a dynamic technological base, moving from 'armchair theorizing', through laboratory and field empiricism, physical hardware models and electrical analogues, to mathematical approaches and computer simulation. The list would seem a chronological development, but this is only partly the case, particularly as modern computer-based models will most likely incorporate aspects of, or data derived from, all types – as demonstrated in the pages of *Computer Simulation in Physical Geography* (Kirkby et al., 1992), or in *Computerised Environmental Modelling* (Hardisty et al., 1993).

So whilst modelling has followed the developing trends of geomorphology – moving from qualitative description through to the dual attempts to relate process with form and to discover the ultimate explanations so sought after in reductionist approaches – models have also aided in framing a variety of philosophical bases to this particular geographical science. Good examples are introduced and debated in *Space and Time in Geomorphology*, edited by Thorn (1982), Thornes' (1983) discussion of multiple stable and unstable equilibrium states and *The Scientific Nature of Geomorphology* (Rhoads and Thorn 1996). Additionally, ideas about 'thresholds' and 'landscape sensitivity' concepts have been suggested (Brunsden and Thornes, 1979) and enlarged upon (Thomas and Allison (eds.), 1993; Thomas and Simpson (eds.), 2001).

One key point that has emerged is that there has been a greater realization that the coupled and linear relationships established by regression analysis of processes are no longer an adequate explanation in the earth sciences (Williams, 1983). Different forms of distribution prevail, requiring differing modelling approaches. In a way, unstable or multiple-stable equilibria make predictive modelling a much more difficult task, reflecting perhaps the ideas propounded by chaos theory, 'whereby fully deterministic relationships appear to lead to apparently random results' (Hardisty et al., 1993) or in the evocative description given by Gleick (1988), that:

> *quite simple mathematical equations could model systems every bit as violent as a waterfall. Tiny differences in input could quickly become overwhelming differences in output ... In weather, for example, this translates into ... the Butterfly Effect – the notion that a butterfly stirring the air today in Peking can transform storm systems next month in New York.*

Such non-linear notions have been further explored by Phillips (1995, 1999a, b), whilst higher scale resolution in both space and time has occupied others. As Bates and Lane (2000) comment, despite enhanced data methodologies '... a model is only as good as the theoretical description that it embodies ...' Continuing also are problems associated with the dimensional shift from 2D to 3D, and the validation, calibration and parameterization of models – all of which raise new research questions and challenges.

1.6 Reflections on Quaternary studies

> *The limits of ice action. Ups and downs of land and sea. Dating and correlation – new methods, old problems.*

Studies of the subdivision of geological time known as the Quaternary owe their origin to the development of ideas about glacial processes and landscapes which emerged in the 1840s, from work in the Alps undertaken by Louis Agassiz, who is accredited with the 'revolutionary' concept on the effects of the formation of ice and its subsequent movement across the landscape. Agassiz had been invited to Britain and, after a series of field observations, he concluded (1838) that all of the country had previously been beneath an ice-sheet and that the so-called 'drift' (accumulations of loose sediments) could be attributed to the action of ice (rather than the effects of some catastrophic flood or marine submergence) as glaciers moved across the landscape – moulding and subsequently depositing eroded material as they passed.

Such ideas were slow to take hold in Britain, despite the widespread field evidence and the fact that current analogues were available in Europe. The theory propounded by Agassiz also accorded with the newly established ideas of Charles Lyell on uniformitarianism. However, high sea levels and marine submergence were still thought to have given rise to the 'shelly' tills and sandy deposits found in mountainous areas such as North Wales (Lewis, 1970). These interpretations prevailed until around the turn of the century, when ideas were put forward about glacial scouring of sea-floor areas exposed by falling sea-levels consequent upon the mass transfer of water in its frozen state onto the land.

Once established as a verifiable theory, glacial erosion and deposition were to foster a long-term interest in the forms and mechanisms of ice erosion, but also into both the nature and extent of former ice cover as well as the timing and number of glacial advances. From these beginnings emerged some major sub-disciplines within the geographical sciences: glaciology, with its dual concentration on the formation and physical properties of moving ice (see Paterson, 1994) and process/form relationships (see Benn and Evans, 1998); geocryology, which addressed frozen ground phenomena and landforms in periglacial environments (see French, 1976 and Washburn, 1979); and a more diverse group of researchers (known collectively as Quaternary scientists) drawn from biology, geology, archaeology, climatology, geomorphology and biogeography, who now form multidisciplinary teams investigating palaeo-environments.

The literature is voluminous and intricate in its analysis, and somewhat bedevilled in the jargon surrounding a necessary desire to correlate events not only between differing parts of Britain and Ireland, but also with Europe and Scandinavia, North America and elsewhere. A 'flavour' of such output can be perused in West (1963, 1968), Lewis (1970), Sparks and West (1972), Mitchell et al. (1973), Bowen (1978), Williams et al. (1993), Ehlers et al. (1991) and in the Geological Society's special revised correlation report (Bowen (ed.), 1999).

The Quaternary has been characterized by significant changes in climate, so that although the effects of ice on landforms and the extent of ice coverage was a predominant early occupation, later focus was on interglacial episodes – especially the biotic assemblages whose materials provided such a rich medium, enabling the important shift from 'relative' dating (utilizing the basic law of superimposition) to the more reliable 'absolute' chronologies of changing environmental conditions provided by the plethora of dating techniques that have been developed in the last 30 years or so (see, for example, Bowen, 1978 and Lowe and Walker, 1997). One of the key points to emerge from the considerable concentration of effort by Quaternary scientists on the dating front is that, despite the advance and sophistication of modern dating techniques and associated research methodologies, there is still some uncertainty surrounding the exact nature and magnitude of environmental change in terms of the processes involved and the resulting landforms in both spatial and temporal contexts. Goudie (1977) gives an indication of the problems, as do Dawson (1992) and Roberts (1994, 1998).

Paralleling the stratigraphy and dating debates has been a concern with the effects of changing sea levels. Dury (1959), in his popular introductory text on geomorphology, coined the phrase 'Ups and Downs of Land and Sea', which set the scene for Stephens and Synge's (1966) more detailed discussion of 'Pleistocene Shorelines'. Glacio-eustasy and isostatic rebound are the process mechanisms, whilst raised beaches, re-emergent submerged forests, trimmed coastal solifluction deposits, re-excavated wave-cut platforms (with or without giant erratic boulders) and buried channels are the local landscape expressions of these changes. Work on sea levels and isostasy was presented by Smith and Dawson (1983), whilst Tooley and Shennan (1987) and Tooley and Jelgersma (1992) have produced edited texts which further develop the sea-level change theme.

As Oldfield (1987) commented, 'No geographer trying to understand the physical landscape of the present can escape a deep concern for its past'. Yet in many ways, the detailed investigations of Quaternary deposits and environments, aided by new methodologies, have turned Lyell's uniformitarianism 'the present is the key to the past' mantra, into 'the past is the key to the future'! Certainly, much of the current research on environmental change (which is of so much concern as we get to grips with human influences on earth systems) is informed by the diverse and prolific studies in the wide-ranging and multi-disciplinary field of Quaternary science.

1.7 Application and environmental awareness

Physical and human geography shake hands again! Integration everywhere. Managing sustainable futures. Scientific uncertainty versus the Decision Makers.

One of the most significant trends which characterizes physical geography over the last 25 years or so has been a concern for the effects of humans on the environment. This has manifested itself in two key ways: first, and derived from a desire to understand more clearly complex process interactions, the monitoring of change across the full suite of natural systems, and second, a concern about human influences and detrimental modifications to these systems. Consideration of the latter has lead quite clearly to the need to manage the environment more effectively, thus sustaining a viable future for us all. Allied with this have been renewed interests in sub-disciplines such as applied geomorphology and natural hazards.

Human interaction with the environment is not a new theme, as Sherlock's book *Man's Influence on the Earth* (1931) shows. His concerns were mainly related to the quarrying of geological materials, although there were interesting sections devoted to coastal change, the circulation of water and an early anxiety about climate change due to carbon dioxide emissions from the burning of fossil fuel. Sherlock concluded that population increase was the main culprit – a thesis which was to be more fully explored in the 1970s (Ehrlich and Ehrlich, 1972, amongst others). Similarly, the human effects on the environment treatise had been extensively reviewed well before the 1970s, in *Man's Role in Changing the Face of the Earth* (Thomas (ed.), 1956).

It is difficult to pick out the particular branch of physical geography which most clearly identifies with these early concerns about the environment, although work on land clearance, poor land management and subsequent soil erosion probably figured most significantly (see, for example, Jacks and Whyte, 1939). However, rainfall–run-off relationships, the effects of changing land use in river catchments, the hydrological basis for water supply and water quality/pollution issues have clearly been a focus for a considerable output of literature on the human influence front. Samples are provided by *Water, Earth and Man* (Chorley, 1969), *Geomorphology in Environmental Management* (Cooke and Doornkamp, 1974), *Water in Environmental Planning* (Dunne and Leopold, 1978), *Man's Impact on the Hydrological Cycle in the United Kingdom* (Hollis, 1979),

Human Activity and Environmental Processes (Gregory and Walling, 1987), *Land, Water and Development* (Newson, 1992) and the five editions of *The Human Impact on the Natural Environment (Goudie, 2000)* – although in some of these texts, the approach is wider than just fluvial systems. Of additional interest (at least in terms of water out of control and perceived risk) is Ward's *Floods: A Geographical Perspective* (1978) and Smith and Ward's *Floods: Physical Processes and Human Impacts* (1998).

Many of these approaches show how geomorphology has concerned itself, through applied studies, with human interactions with landscape changes (Sims, 1990), and how the subject is also able to inform on the errors of the positioning of settlement in certain locations and of methods or approaches which might help to minimize such risks. The high incidence of urbanized floodplain flooding during the winters of 1999–2000 and 2000–2001 illustrate this, as do the concerns about global warming and the threat to coastal communities from sea-level rise and coastal erosion (Viles and Spencer, 1995) or the need for coastal and estuarine management (French, 1997). In this sense, physical and human geography are brought closer together – sharing the ideal of effective sustainable environmental management. The involvement of geomorphologists (Hooke, 1988) and other physical geographers in a range of environmental issues and planning (for example, in catchment plans, integrated river basin management, shoreline management plans and attempts at more complex integrated coastal zone management) reinforces the notion that a shift has occurred linking pure with applied research and perhaps forging closer links again with colleagues in human geography.

Aiding these developments are a number of relatively new methods of environmental monitoring. Remote sensing, from both satellite and airborne sensors (Foody and Curran (eds.), 1994; Barrett and Curtis, 1999), the use of Geographic Information Systems (GIS) for digital data storage and manipulation (Atkinson and Tate (eds.), 1999), databases and metadata for modelling are all examples. A very recent trend is to combine these new technologies into the building of 'expert' systems, utilized as a decision support methodology in environmental management (Moore, 2000).

This brings me to a final point, which relates to the nature of the material that physical geographers provide about environmental processes. Sometimes, the nature of the information base is extremely diverse and often in a form which is totally indigestible by non-specialists (Sims, 1998). The quite natural uncertainties inherent in the scientific approach lead to frustration in decision-makers, who are usually under considerable pressure to 'do something' as a response to a high magnitude–low frequency event. There is a need to communicate ideas and come up with answers from which the element of uncertainty has been removed. This remains a somewhat elusive goal in scientific environmental management.

1.8 Conclusion

So, where have we come from and where are we going to, and does it matter? In the sense that geography is what geographers do, probably not! However, this over-simplification or even a refusal to accept a philosophical base will not do. It is fundamentally important that physical geographers recognize and understand the waypoint markers that have been passed in fashioning the sub-discipline. Knowledge, like landscape, is polygenetic, and whilst physical geographers 'graze' and hopefully absorb some of the nutrients from cognate disciplines, the 'foodstuff' needs to be fit for the purpose intended. Decision on that fitness can only be informed by an acquaintance with what has gone before.

This chapter has therefore attempted not only to provide some historical context to modern research into landscape processes, but also to identify a number of themes which have been of recurring interest throughout the evolving philosophies of geomorphology. I am only too conscious of the very large number of gaps in the account – the result of the enormity of the task and the limited space available. Equally, there is much I would have liked to address from other fields of physical geography, such as biogeography or climatology, but even in geomorphology there are some significant omissions – for example, arid and karst landscapes.

The most striking impression is the health of the subject, measured not only by the sheer volume of published output over the last four decades, but also in the vitality of the subject matter and in the willingness displayed by researchers to explore new initiatives as well as the variations on more traditional themes. Long may this continue in either the reductionist or holistic approaches to geomorphology, although I have to admit to a preference for the latter.

References

Agassiz, L. 1838: 'On glaciers and the evidence of their having once existed in Scotland, Ireland and England'. *Proceedings of the Geological Society of London* 3, 327–332.

Atkinson, P.M. and Tate N.J. (eds.) 1999: *Advances in Remote Sensing and GIS Analysis.* John Wiley, Chichester.

Barrett, E.C. and Curtis, L.F. 1999: *Introduction to Environmental Remote Sensing,* 4th edition. Stanley Thornes, Cheltenham.

Bates, P.D. and Lane, S.N. (eds.) 2000: *High Resolution Flow Modelling in Hydrology and Geomorphology.* John Wiley, Chichester.

Benn, D.I. and Evans, D.J.A. 1998: *Glaciers and Glaciation.* Arnold, London.

Bennett, R.J. and Chorley, R.J. 1978: *Environmental Systems: Philosophy, Analysis and Control.* Methuen, London.

Berry, L. and Ruxton, B.P. 1959: 'Notes on weathering zones and soils on granite rocks in two tropical regions'. *Journal of Soil Science* 10, 54–63.

Board, C. 1967: 'Maps as Models'. In R.J. Chorley and P. Haggett (eds.), *Models in Geography.* Methuen, London, 671–725.

Bowen, D.Q. 1978: *Quaternary Geology: A Stratigraphic Framework for Multidisciplinary Work.* Pergamon, Oxford.

Bowen, D.Q. 1999: *A Revised Correlation of Quaternary Deposits in the British Isles.* Geological Society, London, Special Report No. 23.

Brown, E.H. and Waters, R.S. (eds.) 1974: *Progress in Geomorphology.* Institute of British Geographers London, Special Publication No. 7.

Brunsden, D. 1964: 'The origin of decomposed granite on Dartmoor'. In I.G. Simmons (ed.) *Dartmoor Essays*, The Devonshire Association, Torquay, 97–116.

Brunsden, D. (ed.) 1971: *Slopes, Form and Process*. Institute of British Geographers London, Special Publication No. 3.

Brunsden, D. and Thornes, J.B. 1979: 'Landscape sensitivity and change'. *Transctions of the Institute of British Geographers* NS 4, 463–484.

Budel, J. 1963: 'Klima-genetische Geomorphologie'. *Geographische Rundschau* 15, 269–285.

Burt, T.P. and Walling, D.E. (eds.) 1984: *Catchment Experiments in Fluvial Geomorphology*. Geobooks, Norwich.

Carson, M.A. and Kirkby, M.J. 1972: *Hillslope Form and Process*. Cambridge University Press, Cambridge.

Chorley, R.J. 1957: 'Climate & morphometry'. *Journal of Geology* 65, 628–638.

Chorley, R.J. 1962: 'Geomorphology and general systems theory'. *US Geological Survey Professional Paper* 500-B, 1–10.

Chorley, R.J. 1969a: 'The drainage basin as the fundamental geomorphic unit'. In R.J. Chorley (ed.) *Water, Earth and Man*. Methuen, London, 77–100.

Chorley, R.J. 1969b: *Water, Earth and Man*. Methuen, London.

Chorley, R.J., Beckinsale, R.P. and Dunn, A.J. 1973: *The History of the Study of Landforms, Vol. II, The Life and Work of William Morris Davis*. Methuen, London.

Chorley, R.J. and Kennedy, B.A. 1971: *Physical Geography: A Systems Approach*. Prentice Hall, London.

Church, M. 1980: 'Records of recent geomorphological events'. In Cullingford, R.A., Davidson, D.A. and Lewin, J. *Timescales in Geomorphology*. John Wiley, Chichester, 13–29.

Clayton, K.M. 1964: *A Bibliography of British Geomorphology*. Philip, London.

Clements, F.E. 1916: *Plant Succession, an Analysis of the Development of Vegetation*. Carnegie Institute, Publication No. 242, Washington DC.

Clements, F.E. 1928: *Plant succession and indicators*. H.W. Wilson, New York.

Conacher, A.J. 1988: The geomorphic significance of process measurements in an ancient landscape. In A.M. Harvey and M. Sala (eds) *Geomorphic processes in environments with strong seasonal contrasts, Vol 2: Geomorphic systems, Catena* Supplement 13, 147–164.

Cooke, R.U. and Doornkamp, J.C. 1974: *Geomorphology in Environmental Management*. Oxford University Press, Oxford.

Cooke, R.U. and Warren, A. 1973: *Geomorphology in Deserts*. Batsford, London.

Corbel, J. 1959a: 'Vitesse de l'erosion'. *Zeichschrift fur Geomorphologie* 3, 1–28.

Corbel, J. 1959b: 'Erosion en terraine calcaire'. *Annales de Geographie* 68, 97–120.

Cotton, C.A. 1941: *Landscape as Developed by the Processes of Normal Erosion*. Cambridge University Press, Cambridge.

Cotton, C.A. 1942: *Geomorphology: An Introduction to the Study of Landforms*. Whitcombe and Tombs, Christchurch NZ.

Dana, J.D. 1879: *Manual of Mineralogy and Lithology*, 3rd edition. Trubner and Co., London.

Darwin, C. 1842: *The Structure and Distribution of Coral Reefs*. Smith, Elder and Co., London.

Darwin, C. 1859: *Origin of Species*. John Murray, London.

Davies, G.L. 1968: *The Earth in Decay: A History of British Geomorphology* 1578–1878. Macdonald, London.

Davis, W.M. 1909: *Geographical Essays,* edited by D.W. Johnson, republished 1954. Dover Publications, New York.

Dawson, A.S. 1992: *Ice Age Earth: Late Quaternary Geology and Climate.* Routledge, London.

Derbyshire, E., Gregory, K.J. and Hails, J.R. 1979: *Geomorphological Processes.* Dawson, Folkestone.

Doornkamp, J.C. and King, C.A.M. 1971: *Numerical Analysis in Geomorphology.* Arnold, London.

Douglas, I. 1967a: 'Man, vegetation and the sediment yield of rivers'. *Nature* 215, 925–928.

Douglas, I. 1967b: 'Erosion of granite terrains under tropical rain forest in Australia, Malaysia and Singapore'. *Publications of the International Association of Scientific Hydrology* 75, 31–40.

Dunne, T. and Leopold, L.B. 1978: *Water in Environmental Planning.* Freeman, San Francisco.

Dury, G. 1959: *The Face of the Earth.* Penguin Books, Harmondsworth.

Eden, M.J. and Green, C.P. 1971: 'Some aspects of granite weathering and tor formation on Dartmoor, England'. *Geografiska Annaler* 53A, 92–99.

Ehlers, J., Gibbard, P. and Rose, J. (eds.) 1991: *Glacial Deposits of Britain and Ireland.* Balkema, Rotterdam.

Ehrlich, P.R. and Ehrlich, A.H. 1972: *Population, Resources, Environment,* 2nd edition. Freeman, San Francisco.

Embleton, C. and King, C.A.M. 1975: *Glacial Geomorphology.* Arnold, London.

Embleton, C. and Thornes, J.B. 1979: *Process in Geomorphology.* Arnold, London.

Foody, G. and Curran, P. (eds.) 1994: *Environmental Remote Sensing from Regional and Global Scales.* John Wiley, Chichester.

French, H.M. 1976: *The Periglacial Environment.* Longman, London.

French, P.W. 1997: *Coastal and Estuarine Management.* Routledge, London.

Gleick, J. 1988: *Chaos.* Heinemann, London.

Goudie, A.S. 1977: *Environmental Change.* Clarendon Press, Oxford.

Goudie, A.S. 1995: *The Changing Earth: Rates of Geomorphological Processes.* Blackwell, Oxford.

Goudie, A.S. 2000: *The Human Impact on the Natural Environment,* 5th edition. Blackwell, Oxford.

Green, C.P. and Eden, M.J. 1971: 'Gibbsite in the weathered Dartmoor granite, England'. *Geoderma* 6, 315–317.

Gregory, K.J. 2000: *The Changing Nature of Physical Geography.* Arnold, London.

Gregory, K.J. and Walling, D.E. 1973: *Drainage Basin Form and Process.* Arnold, London.

Gregory, K.J. and Walling, D.E. (eds.) 1974: *Fluvial Processes in Instrumented Watersheds.* Institute of British Geographers London, Special Publication No. 6.

Gregory, K.J. and Walling, D.E. (eds.) 1987: *Human Activity and Environmental Processes.* John Wiley, Chichester.

Hack, J.T. 1960: 'Interpretation of erosional topography in humid temperate regions'. *American Journal of Science* 258, 80–97.

Hadley, R.F. and Walling, D.E. (eds.) 1984: *Erosion and Sediment Yield: Some Methods of Measurement and Modelling.* Geobooks, Norwich.

Hardisty, J., Taylor, D.M. and Metcalfe, S.E. 1993: *Computerised Environmental Modelling*. John Wiley, Chichester.

Higgins, C.G. 1975: 'Theories of landscape development: a perspective'. In W.N. Melhorn and R.C. Flemel (eds.), *Theories of Landform Development*. Binghampton State University, Publications in Geomorphology, New York, 1–28.

Hollis, G.E. 1979: *Man's Impact on the Hydrological Cycle in the United Kingdom*. Geobooks, Norwich.

Hooke, J.M. (ed.) 1988: *Geomorphology in Environmental Planning*. John Wiley, Chichester.

Horton, R.E. 1945: 'Erosional development of streams and their drainage basins: a hydrophysical approach to quantitative morphology'. *Bulletin of the Geological Society of America* 56, 275–370.

Hutton, J. 1795: *Theory of the Earth*. William Creech, Edinburgh.

Jacks, G.V. and Whyte, R.O. 1939: *The Rape of the Earth*. Faber and Faber, London.

King, C.A.M. 1972: *Beaches and Coasts,* 2nd edition. Arnold, London.

King, L.C. 1950: 'A study of the world's plainlands: a new approach in geomorphology'. *Quarterly Journal of the Geological Society of London* 106, 101–127.

King, L.C. 1953: 'Canons of landscape evolution'. *Bulletin of the Geological Society of America* C4, 721–752.

Kirkby, M.J. 1987: 'Models in Physical Geography'. In M.J. Clark, K.J. Gregory and A.M. Gurnell *(eds.) Horizons in Physical Geography*. Macmillan, London, 47–61.

Kirkby, M.J., Naden, P.S., Burt, T.P. and Butcher, D.P.: *Computer Simulation in Physical Geography*. John Wiley, Chichester.

Knighton, A.D. 1984: *Fluvial Forms and Processes*. Arnold, London.

Leopold, L.B., Wolman, M.G. and Miller, J.P. 1964: *Fluvial Processes in Geomorphology*. Freeman, San Francisco.

Lewis, C.A. (ed.) 1970: *The Glaciations of Wales and Adjoining Regions*. Longman, London.

Linton, D.L. 1955: 'The problem of tors'. *Geographical Journal* 121, 470–487.

Lowe, J.J. and Walker, M.J.C. 1997: *Reconstructing Quaternary Environments,* 2nd edition. Addison Wesley Longman, Harlow.

Lyell, C. 1830: *Principles of Geology*. John Murray, London.

Mantell, G.A. 1847: *Geological Excursions round the Isle of Wight*, 3rd edition. Bohn, London.

Mitchell, G.F., Penny, L.F., Shotton, F.W. and West, R.G. (eds.) 1973: *A Correlation of Quaternary Deposits in the British Isles*. Geological Society, London, Special Report No. 4, 67–80.

Moore, A.B. 2000: 'Geospatial expert systems'. In S. Openshaw and R.J. Abrahart (eds.), *GeoComputation*. Taylor and Francis, London, 127–159.

Newson, M. 1992: *Land, Water and Development: River Basin Systems and their Sustainable Management*. Routledge, London.

Oldfield, F. 1983: 'The role of magnetic studies in palaeohydrology'. In K.J. Gregory (ed.) *Background to Palaeohydrology*. John Wiley, Chichester, 141–165.

Oldfield, F. 1987: 'The Future of the Past: a perspective on palaeoenvironmental study'. In M.J. Clark, K.J. Gregory and A.M. Gurnell (eds.), *Horizons in Physical Geography*. Macmillan, London, 10–26.

Ollier, C.D. 1969: *Weathering*. Oliver and Boyd, Edinburgh.

Ollier, C.D. 1995: 'Classics in physical geography revisited: King, L.C. 1953, Canons of Landscape Evolution'. *Progress in Physical Geography* 19, 371–377.

Palmer, J. and Neilson, R.A. 1962: 'The origin of granite tors on Dartmoor, Devonshire'. *Proceedings of the Yorkshire Geological Society* 33, 315–340.

Paterson, W.S.B. 1994: *The Physics of Glaciers,* 3rd edition. Pergamon, Oxford.

Peltier, L.C. 1950: 'The geographic cycle in periglacial regions as it is related to climatic geomorphology'. *Annals of the Association of American Geographers* 40, 214–236.

Penck, W. 1924: *Die Morphologische Analyse.* Englehom, Stuttgart.

Pethick, J. 1984: *Introduction to Coastal Geomorphology.* Arnold, London.

Phillips, J.D. 1995: 'Self-organisation and landscape evolution'. *Progress in Physical Geography* 19, 309–321.

Phillips, J.D. 1999a: 'Divergence, convergence and self-organisation in landscapes'. *Annals of the Association of American Geographers* 89, 466–488.

Phillips, J.D. 1999b: *Earth Surface Systems: Complexity, Order and Scale.* Blackwell, Oxford.

Pitty, A.F. (ed.) 1978: *Geographical Approaches to Fluvial processes.* Geobooks, Norwich.

Playfair, J. 1802: *Illustrations of the Huttonian Theory of the Earth*, reprinted 1964. Dover Publications, New York.

Prestwich, J. 1886: *Geology: Chemical, Physical and Stratigraphical.* Clarendon Press, Oxford.

Rhoads, B.L. and Thorn, C.E. (eds.) 1996: *The Scientific Nature of Geomorphology.* John Wiley, Chichester.

Richards, K.S. 1982: *Rivers: Form and Process in Alluvial Channels.* Methuen, London.

Roberts, N. (ed.) 1994: *The Changing Global Environment.* Blackwell, Oxford.

Roberts, N. 1998: *The Holocene: An Environmental History,* 2nd edition. Blackwell, Oxford.

Ruxton, B.P. and Berry, L. 1957: 'Weathering of granite and associated erosional features in Hong Kong'. *Bulletin of the Geological Society of America* 68, 1263–1292.

Ruxton, B.P. and Berry, L. 1959: 'The basal rock surface on weathered granitic rocks'. *Proceeding of the Geologists Association* 70, 285–290.

Savigear, R.A.G. 1952: 'Some observations on slope development in South Wales'. *Transactions of the Institute of British Geographers* 18, 31–52.

Schumm, S.A. 1956: 'Evolution of drainage systems and slopes in badlands at Perth Amboy, New Jersey'. *Bulletin of the Geological Society of America* 67, 597–646.

Sherlock, R.L. 1931: *Man's Influence on the Earth.* Butterworth, London.

Sims, P.C. 1990: 'Changing Landscapes'. In N. Stephens (ed.) *Natural Landscapes of Britain from the Air.* Cambridge University Press, Cambridge. 232–268.

Sims, P.C. 1998: 'Coastline erosion, protection and management in Devon and Cornwall'. In M. Blacksell, J. Matthews and P.C. Sims (eds.), *Environmental Management and Change in Plymouth and the South West.* University of Plymouth, Plymouth, 73–92.

Small, R.J. 1970: *The Study of Landforms.* Cambridge University Press, Cambridge.

Smith, D.E. and Dawson, A.G. (eds.) 1983: *Shorelines and Isostasy.* Academic Press, London.

Smith, K. and Ward, R.C. 1998: *Floods: Physical Processes and Human Impacts.* John Wiley, Chichester.

Sparks, B.W. 1972: *Geomorphology,* 2nd edition. Longman, London.

Sparks, B.W. and West, R.G. 1972: *The Ice Age in Britain.* Methuen, London.

Stephens, N. and Synge, F.M. 1966: 'Pleistocene Shorelines'. In G. Dury (ed.), *Essays in Geomorphology.* Heinemann, London, 1–51.

Stoddart, D.R. 1967: 'Organism and ecosystem as geographical models'. In R.J. Chorley and P. Haggett (eds.), *Models in Geography.* Methuen, London, 511–548.

Stoddart, D.R. 1969: 'Climatic geomorphology: review and assessment'. *Progress in Geography* 1, 160–222.

Strahler, A.N. 1950a: 'Davis' concept of slope development viewed in the light of recent quantitative investigation'. *Annals of the Association of American Geographers* 40, 209–213.

Strahler, A.N. 1950b: 'Equilibrium theory of erosional slopes approached by frequency distribution analysis'. *American Journal of Science* 248, 673–696 and 800–814.

Strahler, A.N. 1952: 'Dynamic basis of geomorphology'. *Bulletin of the Geological Society of America* 63, 923–937.

Sugden, D.E. and John, B.S. 1976: *Glaciers and Landscape*. Arnold, London.

Thomas, D.S.G. and Allison, R.J. (eds.) 1993: *Landscape Sensitivity*. John Wiley, Chichester.

Thomas, M.F. 1965: 'Some aspects of the geomorphology of domes and tors in Nigeria'. *Zeichshrift fur Geomorphologie* 9, 63–81.

Thomas, M.F. 1966: 'Some geomorphological implications of deep weathering patterns in crystalline rocks in Nigeria'. *Transactions of the Institute of British Geographers* 40, 173–193.

Thomas, M.F. 1974: 'Granite landforms, a review of some recurrent problems of interpretation'. In E.H. Brown and R.S. Waters (eds.), *Progress in Geomorphology*. Institute of British Geographers, Special Publication No 7, 13–38.

Thomas, M.F. and Simpson, I.A. (eds.) 2001: *Landscape Sensitivity: Principles and Applications in Northern Cool Temperate Environments. Catena.* Special Issue 42, 81–386.

Thomas, W.L. (ed.) 1956: *Man's Role in Changing the Face of the Earth*. University of Chicago Press, Chicago.

Thorn, C.E. (ed.) 1982: *Space and Time in Geomorphology*. Allen and Unwin, London.

Thornbury, W.D. 1954: *Principles of Geomorphology*. John Wiley, New York.

Thornes, J.B. 1983: 'Evolutionary geomorphology'. *Geography* 68, 225–235.

Tooley, M.J. and Jelgersma, S. (eds.) 1992: *Impacts of Sea-level Rise on European Coastal Lowlands*. Blackwell, Oxford.

Tooley, M.J. and Shennan, I. (eds.) 1987: *Sea-level Changes*. Blackwell, Oxford.

Tricart, J. and Cailleux, A. 1972: *Introduction to Climatic Geomorphology*, translated by C.J.K. De Jonge. Longman, London.

Twidale, C.R. and Lageat, Y. 1994: 'Climatic geomorphology: a critique'. *Progress in Physical Geography* 18, 319–334.

Viles, H. and Spencer, T. 1995: *Coastal Problems: Geomorphology, Ecology and Society at the Coast*. Arnold, London.

Von Engeln, O.D. 1942: *Geomorphology: Systematic and Regional*. Macmillan, New York.

Walling, D.E. 1987: 'Hydrological and fluvial processes: revolution and evolution'. In M.J. Clark, K.J. Gregory and A.M. Gurnell (eds.), *Horizons in Physical Geography*. Macmillan, London, 106–120.

Walling, D.E. and Webb, B.W. 1983: 'Patterns of sediment yield'. In K.J. Gregory (ed.), *Background to Palaeohydrology*. John Wiley, Chichester, 69–100.

Ward, R.C. 1967: *Principles of Hydrology*. McGraw Hill, London.

Ward, R.C. 1978: *Floods: A Geographical Perspective*. Macmillan, London.

Washburn, A.L. 1979: *Geocryology*. Arnold, London.

Waters, R.S. 1958: 'Morphological mapping'. *Geography* 43, 10–17.

Waters, R.S. 1964: 'The Pleistocene legacy to the geomorphology of Dartmoor'. In I.G. Simmons (ed.), *Dartmoor Essays*. The Devonshire Association, Torquay, 39–57.

West, R.G. 1963: 'Problems of the British Quaternary'. *Proceedings of the Geologists Association* 74,147–186.

West, R.G. 1968: *Pleistocene Geology and Biology*. Longman, London.

Williams, G.P. 1983: 'Improper use of regression equations in the earth sciences'. *Geology* 11, 195–197.

Williams, M.A.J., Dunkerley, D.L., De Decker, P., Kershaw, A.P. and Stokes, T.J. 1993: *Quaternary Environments*. Arnold, London.

Wise, S.M. 1980: 'Caesium-137 and Lead-210: a review of the techniques and some applications in geomorphology'. In Cullingford, R.A., Davidson, D.A. and Lewin, J., *Timescales in Geomorphology*. John Wiley, Chichester, 109–127.

Wolman, M.G. and Miller, J.P. 1960: 'Magnitude and frequency of forces in geomorphic processes'. *Journal of Geology* 68, 54–74.

Wood, A. 1942: 'The development of hillside slopes'. *Proceedings of the Geologists Association* 53, 128–140.

Wooldridge, S.W. and Linton, D.L. 1939: *Structure, Surface and Drainage in South East England*. Philip, London.

Wooldridge, S.W. and Morgan, R.S. 1937: *The Physical Basis of Geography: An Outline of Geomorphology*. Longman, London.

Young, A. 1972: *Slopes*. Oliver and Boyd, Edinburgh.

2

Meaning, knowledge, constructs and fieldwork in physical geography

Stephen Trudgill

2.1 On being there

Poulnagollum Cave – Main Junction
Steve Trudgill, July 1967.
Quiet dark talk in the ripple over stones
eerie glow, listen to the talk of the waters.
Huge caverns beneath the earth
Vast with festoon needles
surging water against my legs
rush and crash, blunder and deafen
force back, slip down in pool, roar
and water all over and force.
Now quiet and warm, sit and eat and listen companions with and water talk
over stones ripple broad brown waters
spread, white fleck.
Beneath the earth, reflect and ponder.
Listen, drip and splash and drip and drip in pool.
Quiet, dark, ripple over stones,
glow, and listen to the talk of the waters
and ever drip to pass the time.
Sharp drop note as caverns build
little by little, caverns build.
Time is from long, long ago
and time is for ever long
and time is for ever
and time is never
for time is *now*.

Self-analysis can be a dodgy business, but I had set myself to find an answer to the question of why physical geography mattered to me. What is revealing is that I rapidly passed by all the texts, writings and scientific analysis and reached for a poem, reproduced above, which I wrote in 1967 about caving while undertaking my undergraduate dissertation on the Burren in Ireland. The elements of force, time and companionship are evident – above all, the feeling of being there, of *experience*. Ask physical geographers why they enjoy their work and many will give you the same kind of answer – a love of wild places, mountains, glaciers, coasts and so on, as a prime motivation.

This much is shared by many people and not unique to physical geographers, however. For the subject of physical geography as a science, I am not sure whether it is the experience which provides the justification, with the science as an extra, or whether the science provides the justification and the experience is the extra. To me that does not matter as they are both providing separate but closely related narratives about the world around us. There has always been something special about working on a rocky shoreline in the teeth of a gale and coming away with a set of measurements to six decimal places. To be sure, the science is an enriching part of the experience. That is to say, that the depth of understanding increases the experience in that you also have the feeling of knowing what is going on around you. You can, of course, have the experience without the science, but that is all the shallower. Equally, the science without the experience is all the hollower. It is thus revealing that I seem to see fieldwork as central to physical geography.

But what of the realm of ideas and concepts? What of those constructs that guide us and which we test reflexively against the evidence and which give us the feeling that we know and understand what is going on around us? What about the actual *meanings* in physical geography?

Evidently, we have already identified experiential meanings – the experience of the force of waters while caving and the sense of time give a sense of meaning – and these merge with conceptual meanings about how we think the world works. How do we derive these senses of meaning?

2.2 Meanings in physical geography

I recently asked a group of undergraduates to look at a series of coastal limestone landforms in Mallorca (Figure 2.1) and then to describe and explain it to me. The limestone rocks were dissected and rugged and, to my mind, etched by biochemical processes involving boring algae and salt weathering (Figure 2.2). Not having researched intertidal karst as I had (I had not yet lectured to them on the subject), the students came back with descriptive words like 'lunar' and they asked if it was not obviously volcanic lava.

Now what was I to do? How legitimate were their deductions? Clearly, I could play 'twenty questions' till they came up with the 'right' answer; or I could just tell them my answer and/or I could set them measuring and making closer observations till they came up with the answer I was expecting. The occasion set me thinking.

In fact, they were not equipped to come up with the 'right answer' and were using the only constructs they had in their mind, in a clearly comparative sense – 'it looks like . . .'

Figure 2.1 Students seeing a Mediterranean limestone coast with landforms they have not studied before.

– something else they already had knowledge of. If they had looked long enough and used hand lenses, eventually they would have discovered the minute pitting which the boring algae leave and, using an acid bottle, they would come up with limestone rather than volcanic rock and so on. But to me this illustrated a general theme: our explanations refer to what we already know, and when presented with the unknown we readily refer to our existing constructs, partly because that is the obvious thing to do and partly as it excuses us somewhat from the effort of trying to see what is actually there. But what, in fact, is actually there?

The questions that arise are ones of how far do we 'force' what is there into existing constructs and how far do we make 'discoveries', leading to new constructs. To me, the answer is perhaps the former first, followed by the latter if the former does not serve us well – in other words, if what we already know does not explain what we observe. However, we have to realize that the pre-existing constructs are not serving us well before we move to new ideas. Thus, my task was to point out features inconsistent with their explanations, but still the task is to get the students to observe – to actually see – rather than rest on their preconceptions.

This is admirably expressed by Bishop (1992) in the phrase 'images do not raise consciousness, they force it to descend'. Just as we might go on holiday abroad and read the guidebooks to an area, we then spend time finding what we have learnt to expect from prior reading. Satisfaction is expressed when we find 'the sights' the guidebooks

Figure 2.2 ... *surely it is volcanic lava?*

have preconditioned us to expect. Actually seeing is very confusing and we need to make a construct of what we see, which is necessarily selective. The selectivity is fundamental and we have to establish the basis of the selectivity. On field classes we often spend considerable effort on finding sites where 'it works' and you can relate a narrative that makes sense, especially when we have become preconditioned to a narrative through textbooks. Coming back with a graph, transect, section, sequence or some other form of pattern is always more satisfactory than saying 'I don't know' and a

Figure 2.3 *Reading the landscape: a coastal scene bathed in sunlight with greens and browns, an eroded coastal platform and a periglacial slope.*

scatter of points on a graph with no evident logic behind it. We inevitably search for meanings, so we tend to look for classic sites where textbooks and observations agree.

Recent writings have suggested that 'information does not just exist "out there" waiting to be collected or "gathered" but is constructed, or created, in specific ... contexts for particular purposes' (Mosse, 1994, quoted from Twyman et al., 1999). In addition, we 'can transform our concept of the field from *a space in which we interpret and read the world*, others objects, etc., to *a space through which the world is produced*. In other words, the field is not a point of access or gateway through which we read and represent "the other". Rather it is the fundamental process through which our own geographic world (as a meaningful narrative) is constituted' (Rose, 1999).

This much is crystallized by Paul Feyerabend in *Conquest of Abundance: A Tale of Abstraction Versus the Richness of Being* (1999). He writes: 'Variety disappears when subjected to scholarly analysis. ... Anyone who tries to make sense of a puzzling sequence of events ... is forced to introduce ideas that are not in the events themselves but put them in perspective. There is no escape: *understanding a subject means transforming it* ... – inserting it into a model or a theory or ... account.' Thus the students thought they were standing on solidified lava because that was the only way they could make sense of it using the means at their disposal.

So is experience a better guide to understanding the world? Are concepts and models a tyranny of imposition which disempowers individual observation? Should I ask students to write poems rather than draw graphs? Or are models and concepts a provisional means of understanding which then have to be subject to verification through a test – which needs to be a fair test (and how do you ensure that it is a fair test?)? In other words, if someone's observation does not fit the preconceptions, are they wrong? Do we then conclude that the observation is erroneous, or that the test wasn't a fair one, or just ignore it because it doesn't fit? This means keeping the model/narrative/construct and disempowering the individual observation. Alternatively, do we reject the model/narrative/construct and empower the individual observation which doesn't fit?

These questions are at the heart of progress in science because things which don't fit our existing concepts obviously lead to new concepts (provided we can be sure that the observation is not erroneous). Or, more fundamentally, is it that the models, narratives and constructs are just that and the world isn't actually like that but nevertheless we cling onto them as they make sense of the world. So, are our constructs a *complete* illusion, which make us ignore things which don't fit – an imposition – or a provisional statement which we readily throw over if the evidence contradicts – a discovery?

My answer is neither – the constructs, narrative and models are provisional statements which, if we are not careful to be aware of their interim nature and social construction, make us ignore things which don't fit. They are working statements by which we arrange things that are neither the truth nor an illusion but are *justifiable*. As such they found a working basis, nothing less, but also, crucially, nothing more. What then is important is not to let them constrain our experience, observation or further thought. There are many ways of seeing things, but we usually choose to legitimize some and marginalize others according to preconceptions.

Clearly, to me, we transform the subject in terms of significance and meaning – what Sack (1990) terms the 'realm of meaning' is clearly crucial. Significance and meaning, then, are derived from our education – the texts if you like – which is really a kind of brainwashing. We see what our constructs (or tutors) tell us to look for. Take a group of students to a cliff top looking at a coast (Figure 2.3). They will see the sea, cliffs, perhaps though memories of holidays, and/or greenness, wetness or whatever, according to the autobiographies and experiences. After the tutor has finished they will see a wave-cut notch, a periglacial slope, raised shorelines, marine plantation surfaces and so on. How the experience has been enriched by understanding and how the subject has been transformed! They may also see a system, which cascades material from the upper coast, through the slope to the shore and beach, and a coast under threat from erosion, with a significance in human terms, like property in danger. What is wrong with that? Nothing, except that once you have given that meaning and significance then they stick – reinforced through the authority of syllabus and marking – and are not easily put aside. What has been given then becomes the reality which can forever constrain the ways of looking at the scene and it takes a strong mind to change the view and present different meanings. Nothing wrong in that, particularly, but when we see what we are looking for (is that actually all we can really do?), is the science really going to progress? The answer is yes if we realize that the **given meaning** is just that and not the only possible meaning. Discovery is, then, seen as finding a new meaning. After all, 'discovering' the workings of nature is really something which makes sense to us – as Hume (1711–1776) argued, 'necessity is something that exists in the mind, not in objects'.

If, then, finding new meanings is to be equated with progression in science, it should be useful to look at different sources of meanings. Perhaps already we can distinguish **Experiential, Given** and **Learned Meanings**. In addition, meanings in science vary in their scope:

1. **Personal Meanings** for the observer/investigator (what does my work mean to me?)

2. **Research Meanings** for the specific topic under investigation (what does this mean for, e.g. the evolution of limestone landforms or forest succession?)

3. **Subject Meanings** for a science as a whole (what does the research mean for geomorphology/biogeography/climatology...?)

4. **Public Meanings** for the wider audience beyond the scientific endeavour (what do the findings of science mean in an applied context or in terms of people's views of the world?).

As an example, for the coastal view above the given meanings are derived from (2) to (3) and then given by the teacher to the students so that (3)s become (1)s.

For each, the criteria of significance and interest are used which refer to a relational nexus of values. The first, personal meaning, relates to the motivations, justifications, knowledge, frame of reference, experiences and autobiography of the investigator. The other three are also derived from and judged by the observer, but are also derived from and judged by others, and this in growing numbers of different people when moving from Research to Public Meanings as the scope broadens from specialists to the wider audience.

The situation is reflexive in that meanings derived from Public Meanings can feed down progressively to Subject, Research and Personal Meanings as much as Public Meanings can be derived from Subject, Subject from Research and Research from Personal Meanings.

There is a well-known hiatus between Subject and Public Meanings in terms of the public understanding of science when moving from the former to the latter. There can also be a gulf between them in the other direction in terms of any political direction of Science by Public Meanings.

The relationship between Subject and Research Meanings is a closer one and is often seen in terms of general scientific paradigms in the way the former can influence the latter, often in a way that is not always conscious or explicit. Here we might consider the ways in which accepted or unchallenged paradigms, models or constructs might act to constrain or otherwise influence research. Are we imposing models on the world and/or discovering the world? Research Meanings often implicitly feed into Subject Meanings or, on occasion, challenge them, acting to change the paradigms.

One of the most interesting relationships is, however, between Personal Meanings and Research Meanings. First, the relationships between Personal, Research, Subject and Public Meanings are not necessarily hierarchical, with the necessity of Research and Subject Meanings intervening between Personal and Public Meanings. There can indeed be a direct relationship between the Personal and the Public in that the former might be directly derived from the latter. This then might be used to derive the Research Meaning and influence the Subject Meaning in a 4-1-2-3 manner. The subject is collectively influenced by individuals self-motivated by public concerns.

This raises a whole series of questions of individual motivations for undertaking research which may be derived from a number of sources, not just the general subject paradigms. The actual research undertaken is thus influenced by this reflexive, relational nexus. Involved are not only general scientific approaches in the subject, but also the serendipity which arises from the various sources of personal motivations and from the public impressions of science, which escape the bounds of the subject as orthodoxies, and the varied interactions between the different levels of meanings.

Such a discussion of reflexivity may be unusual in physical geography in that many practitioners, if asked, would cite their motivations (and therefore the meanings they derive from their work) as simply involving finding out how the world works. This involves the search for mechanism, pattern and logic.

There are, however, often the additional spurs, such as liking the location of their work, which may be expressed as 'being outdoors/on glaciers/up mountains'. This may be with the additional plus of being expeditionary, with teamwork and associated comradeship. A preference for wilderness and nature in some form is thus also involved. Often the sense of 'being useful' is also present – to the subject and/or to humanity in general. Overall, however, the sense of 'finding out something' tends to be dominant.

The question of 'to what are you referring?' (i.e. what are your reference criteria, why are you doing it, what is a success?) may be met with puzzlement. The sense of 'fit' tends to prevail – data fit a pattern and a sense of excitement and satisfaction occurs, or data challenges a preconceived notion, leading to a sense of a new discovery. Senses of self-fulfilment are derived from concordance or 'fit' ('it worked') or from discordance ('it must be something else') and satisfaction at finding an alternative explanation. The idea of forcing a construct or model onto the physical world is met with puzzlement if not resentment because the air of 'discovery' with attendant pattern and logic prevails.

Students thus tell me that physical geography is safer (or easier, or, conversely, more boring) than human geography because in the former the interaction between the observer and the observed is not such an issue as it can be in the latter. It seems that there are rules in physical geography: hot air rises and water flows down hill. Thus the physical environment must follow such rules and we only have to discover them and how they interact (or have interacted in the past) to unravel the relationships between cause and effect. Are these impressions true?

The answer lies in the nature of the phenomena being studied and the scale at which one is working. The element of discovery of processes and their relationship with pattern becomes more justifiable at a small scale and in situations where few components interact. Soil particle movement on a bare surface during a rainstorm can be readily related to raindrop size and the associated kinetic energy which gives the potential to dislodge particles. The key factor then involves the varying cohesiveness of particles of different sizes (sand, silt and clay) which, in turn, is influenced by soil moisture and soil organic matter content. Further factors can be built in to calibrate for differing situations where different conditions obtain (e.g. different slopes) and/or different combinations of conditions obtain (varying compositions of organic matters or clays, etc.). Such relationships should be measurable and not subject to the way in which it is measured beyond the technical accuracy and precision of instrumentation. Moreover, and most importantly, several different observers can reproduce the same set of relationships by working in many different ways and with many different sets and levels of background knowledge. In this sense, though there will usually be a personal meaning, if anyone can test something and find it to be true, then this is independent of Personal Meanings.

However, while we might faithfully establish experimentally a series of reproducible observations, this is not the same as answering the question of what happens at a particular place in a particular set of conditions. It is in more complex situations where there are a contingent set of conjunctions that the understanding of a particular situation becomes more related to the questions we ask, what answers they give us and how we interpret the findings. This is when reflexivity between the observer and the observed sets in. Interpretation becomes much more a matter of our personal meanings, significance placed on observations, motivations for the questions we ask, sets and levels of knowledge, models we have in our minds and subject paradigms. Imagine now not just trays of different soils of varying composition and moistures subject to impact by drops of water with varying kinetic energy and the attendant resulting graphs. Imagine a whole hillside with varying degrees of vegetation cover, soils and soil moisture and temporally varying rainfall intensity. Can we predict the sediment yield? Well, yes. We can duplicate our experiments to allow for varying vegetation cover, varying rainfall intensity, varying prior conditions, and so on, to produce an aggregated prediction according to the varying combinations. And we can keep going until we reproduce a verifiable result (assuming that prediction matching observed constitutes an explanation or verification). But you can see where we are going. The next storm will be different and so we have to re-calibrate our prediction for a different combination, and so on. Before long, however, we get into a situation where we either might just as well measure the sediment loss in each and every storm and not bother with prediction, or we simplify the situation and use a simple 'rule of thumb' model that gives a satisfactory prediction if not totally predicting everything (the obvious example being the much maligned but useful in practice Universal Soil Loss Equation, Wischmeier and Smith, 1978).

However, the 'tyranny of models' can set in and we might then believe the models as 'true' and discard things which don't fit. This is what Kearns (1998) terms the virtuous circle of facts and where we have the choice of ignoring the facts and keeping the model or recognizing the facts and modifying the model.

2.3 The nature of knowledge

There are those who see themselves as scientists building up a body of knowledge about how the physical environment works, including the study of the processes and patterns of climates, landforms and vegetation. Then there are others who seek to deconstruct knowledge, rehearsing that all knowledge is positioned relative to the knower, with their association concepts and preconceptions.

The latter approach is sometimes presented as a conclusion and is a point of view that might command an automatic acknowledgement, but then I ask: where does this approach get you? It should at the very least be a starting point rather than a conclusion. The key point, to my mind, is: why do environmental scientists hold particular views of the world? What characteristics and autobiographies of the scientists, and what aspects of the society, times and ideas of their age, have influenced the views of the scientists? I also ask for what aspects of scientific knowledge is this deconstructionist approach appropriate and the most relevant?

In my view, there is a spectrum of knowing. On one level, a tree is a tree and a hillslope is a hillslope. However, a tree can also be seen in the context of a system, with flows of water, energy and nutrients as a component of that system. A hillslope can be seen as a cascade of materials and energy, as part of a cycle of erosion or as evidence of historical processes. There are thus simple concepts ('tree', 'hillslope') and wider meta-concepts ('ecosystem', 'cascade'). These meta-concepts lie somewhere between 'hot' and 'beautiful' – hot is immediately verifiable and beautiful is a value judgement. Concepts such as 'ecosystem' and 'hillslope cascade' are neither immediately verifiable by observation nor a value judgement; they are *supportable constructions*. My feeling is that as one moves from that which is verifiable, through observation, through supportable constructions, to value judgements, the deconstructionist approach becomes increasingly relevant. To recognize a tree one does not often need to know the background and nature of the observer, though occasionally this may be relevant as definitions may also be. To recognize a woodland as an ecosystem needs that construct to be in place in the viewer, and to recognize a woodland as timber, wilderness, attractive or threatening, certainly depends on the nature of the viewer.

Given that even something verifiable may depend on definition, I see four levels of knowledge:

1. independent reliability

2. dependent reliability

3. construct

4. value judgement.

1. A reliability is verifiable by observation and may be:

(a) tangible – verifiable by experience, e.g. hot, cold

(b) intangible – the world is round, the earth goes round the sun.

In general, a tangible independent reliability is inescapable and free of definition or viewpoint. This is the nearest definition of a 'fact' and can be readily related to direct experience. It is difficult to see how such knowledge could be positioned and relative to the investigator except in matters of degree and tolerance through familiarity. There is little or no room for argument.

Intangible independent reliabilities are verifiable but not through direct, everyday experience. They may, however, reflect the age of the times and the progress of discovery and include the overcoming of counter-assertions (the world is flat, the earth is the centre of the universe) and the acceptance of the arguments of others. Such an acceptance amounts to a belief and often also involves the role of 'experts' and may relate to the perception of risk. Such knowledge can thus be positioned within the societal context. Dissension may arise from the lack of first-hand experience which allows counter-beliefs to exist.

2. A dependency is also verifiable by observation and/or measurement, but tends to be dependent upon the observation and/or measurement and also subject to definition. Here we might have a central concept but the limits are not always clear (e.g. tree, but when is a tree a shrub? Some definition, e.g. minimum height, might be needed). In addition, the way measurements are taken may have an influence, in particular the scale of measurement or the position of the measurer. Dependencies may also be tangible or intangible in that we might measure the height of shrubs and trees for ourselves or take an astronomer's word for interplanetary relationships. Dissension arises from measurement perspectives and definitions, which allow different results to be obtained.

3. A construct is a way of looking at the world and is highly dependent upon the viewer, their education, concepts, tenets, theories and perceptions as well as the paradigms of the age. A forest can be seen as an ecosystem, marketable timber, etc. Constructs are inherently intangible or tangible but selectively so, with evidence dependent on perspectives and preconceptions.

4. A judgement depends on values which can be both spiritual and instrumental.

These categories can merge from one to another and, in particular, the distinction between dependencies and constructs can be blurred. Research may progress from reliabilities, through dependencies, to form constructs and lead to value judgements. Equally, existing constructs may influence, constrain or facilitate our investigation of dependencies.

The results of any investigation can be interpreted in the worlds of meaning and significance both in terms of the discussions within a subject and the significance placed on them externally.

In physical geography, reliabilities may or may not have been obvious through history, or may have been discovered at some time, but either way they are now implicit assumptions – water will flow downhill under gravity (unless there is confinement and a hydrostatic head), hot air rises, cold air is denser and sinks, and so on. Thus, some are universal laws and some need the qualification that they only obtain under certain conditions.

Much of what physical geographers do involves dependencies. Much scientific work involves observations which are made, the coincidence of which is taken as an indication of pattern. Conflicting observations give rise to further work. This is classically the case in Quaternary reconstruction, where interpretations of past sequences and the operation of past processes can be site dependent: scientific meaning is derived when observations from a number of sites concurs. Much of process geomorphology depends on measurements and quantification and also the statistical treatment of numerical observations in the search for general patterns. They shift because they are dependent on current observations, and further observations may change the situation.

The distinction between dependencies and constructs is blurred for two reasons. First, collectively dependencies lead to the formation of constructs through generalisation. If observations do not fit with the derived construct then either the construct only applies in particular conditions which do not obtain for the anomalies (the construct is intact) or the construct is inappropriate and has to be revised. We might call these *working constructs*. Second, constructs influence what we investigate, observe and/or measure. We might call these *meta-constructs* which have an overarching influence on the way we go about any investigation of dependencies.

Constructs themselves may thus exist where only fragmentary observations are available, such as in the reconstruction of erosion surfaces. By contrast with depositional sequences, where the evidence has amassed over time, with erosion, by definition, much of what might have been observable has inevitably vanished. In other words, there is very little in the way of verification – observations then become evidence and these constructs almost amount to a belief. This can be challenged by other forms of interpretation and/or of logical argument. The approach is however, aggregative, the construct being the outcome of some existing evidence together with logical argument, and they are an end product. These can be called *interpretative constructs* and differ from working constructs because in the latter there is substantial and substantive evidence. They can then also become *prior constructs.*

Prior constructs can be either working or meta-constructs and are held prior to an investigation and perhaps, thereby, influence the approach and any outcome. Thus they may be derived as working constructs from abundant evidence or as interpretative constructs semi-independently from any evidence. As prior meta-constructs, they can be derived from a philosophical viewpoint and can survive because they do not conflict with any evidence and are otherwise supportable. Models of cycles of erosion and 'the ecosystem' are good examples. These may be termed *prior constructs* which have an important influence on an outcome – and, of course, can be challenged by differing viewpoints but remain inherently untestable. Their utility lies in the way they are valued as organizational concepts and they can be used to marshal reality and, indeed, constrain it. As they are essentially untestable, it is difficult to disprove them, and as they are organizational concepts, support can always be found for them. They can, of course, always be challenged by different organizational constructs.

What interests me is the way we go about fieldwork. We like to profess that we are open minded and can change our constructs in the light of new evidence. However, I suspect that both in field research and in field teaching it is often the other way round – we fit our observations into our constructs. Indeed, it may be deeper than that; we only see what our constructs guide us to see and find significant.

2.4 The fieldwork experience

Our first introduction to fieldwork is usually the field trip where teaching is undertaken. This then sets the scene for later field research and, as such, field teaching has a fundamental influence on field research. What interests me here is the derivation of personal meanings through the field experience and the relationships between personal experiences and given meanings.

Given meanings can represent an enrichment or a replacement of prior personal meanings. In field teaching, an allowed plurality of meanings is preferable to the dominance of a given meaning. The plurality not only refers to multiple knowledges in terms of nature as contested space but also to the range of experiences which involve both the affective emotions as well as cognitive knowledge. Meanings derived in the field which relate to the affective emotions are seen as important, if not more important, than the cognitive meanings derived from imparted knowledge. The field work experience often leaves strong memories of affective emotions which are more life-enriching than the cognitive material alone. Thus, given and personal meanings have elements of competition and coexistence.

The relationships between personal experiences and given meanings can be illustrated by the following spectrum.

1. Enablement of personal meaning through given experience

When I am trying to paint a picture on an art course, say I struggle with perspective and feel that the cottages I am drawing look wrong. The art tutor then talks about vanishing points and gives me a lesson of perspective. After this, I can draw houses receding into the distance in the correct proportion and I feel that they look right. It is still my painting and someone else helped me to achieve my personal goals. My incentive – to draw so it looks right to me – has been retained. My personal meaning is retained and the 'given' aspect relates to how best I can retain my personal meaning – education as enablement.

2. Enrichment through given meaning

When I first walked round Cambridge I saw what I saw – old buildings, market, shops and so on, things that seemed significant to me. When I then went on a tour of Cambridge with a historical geographer, his information greatly increased my sense of significance as I learnt the sequence of events, the signs of changes through the ages, giving a great depth of meaning and a sense of history, which I welcomed. Now as I walk round Cambridge I have a much-enhanced sense of significance round every corner. I acquired given meanings in addition to my own prior meanings – education as enrichment.

3. Conscious acceptance of given meaning

Trudgill, in *Accent, Dialect and the School* (1975) and in other writings, has discussed the case of how, if you are not taught to speak 'properly' (received pronunciation, RP) it is felt that you will not 'get on' in life, let alone get a job in some sophisticated context if you sound like a country bumpkin. There is an incentive of getting on, but a loss of personal meaning in terms of the rich cultural language of your local origins and identity. One personal meaning is replaced by another, but it is consciously undertaken. This may be viewed as culturally regrettable in terms of dialect loss, and while, in some senses, the individual may feel that they have little choice, the given meaning is recognized by the individual as a way of changing personal meanings more advantageously – education as simultaneous disempowerment and enablement.

4. Disempowerment of self through enforced given meaning

In Hugh Brody's *The Other Side of Eden* (2001), there is a representation of education as abuse. Brody records how Inuit people were taken to residential schools and beaten and punished if they used their own language rather than the given English, with the attendant loss of culture and identity. The given meaning is imposed deliberately and the incentive for the individual is zero, apart from avoiding punishment. It is intended that the given meaning will replace the personal meaning – education as imposition.

While there are many nuances between and around these items, now that we have established a spectrum of what I am talking about, we can ask, where do field education and field research education lie in this spectrum?

In field teaching, much of the processes happen in the contexts of meeting the syllabus and the incentive of examinations, which must be no more than a socialization of norms in a way that has been agreed by society, or that part of it consulted, through the syllabus. As such it is defensible, but the given meaning is clearly the 'right' meaning reinforced through the examination process.

I leave the reader to draw what parallels they will between the first and second set of examples, but it should be clear that I am asking to what extent is field teaching an enablement, an enrichment, a difficult to avoid acceptance of given meanings and/or even an abuse?

If given meanings dominate, does this seem like a disempowerment of the individual? If personal meanings dominates, does this seem like truancy? Is the ideal situation one where a reconfigured personal meaning results from the integration of the experiential meaning and the given meaning, i.e. where the given meaning is owned?

When looking at plants, animals and landscapes in the field, does not the teacher emphasize one singular given meaning which has to be rehearsed for examinations? Is this still not the same when the student 'discovers' the given meaning for themselves? Does not the given meaning become the dominant meaning – rather than one of a plurality of allowed meanings? This to me becomes important when looking at the landscape, in that teaching landscape as geomorphology could be giving a dominant meaning. Enriching as this might be, it is only one way of seeing the landscape and in fact there are many different viewpoints involved – 'nature as contested space' (Macnaghten & Urry, 1998). Is this, in fact, disempowerment masquerading as enrichment and does this not happen because of the possible 'tyranny' of the syllabus and examinations? In general terms, if you are always going to quote Darwin, or whatever 'authority', are students ever going to develop their own ideas? This is not to deny the importance of examinable taught meanings, but a plea that they should coexist rather than dominate or replace prior meanings.

There are, therefore, two aspects of the relationships between given meanings and personal meanings: the degree to which they compete and the way in which any competition is resolved (through dominance of one or the other or through coexistence).

Additionally, however, fieldwork is much more than a matter of the placement of given didactic meanings. Irrespective of what is being taught, there is the strong element of the fieldwork experience by which new personal meanings are derived. This involves the physical experience and the closely related social experience – senses of achievement, social interaction and being outdoors evoke strong affective emotions which are important in personal development.

We know that when we learn something, it clearly means more if we can relate it to our experiences, which is why educators often use metaphors in everyday experience to explain new or difficult concepts (equilibrium is like a football in a valley: wherever you kick it, it comes back to rest at the bottom, in contrast to a football on top of a hill). However, when we are field teaching, do we not tend to sideline the commonplace affective emotional experience and stress the taught (syllabus) meaning as the dominant one? I feel that from many points of view, it is important to look at the *whole* fieldwork experience.

Following the four-stage spectrum of empowerment to abuse, above, the spectrum would begin with a Van Matre (1974) approach to immersion and with wilderness as therapy where given meaning involves facilitating your own meanings.

Then move to a 16–19 syllabus which starts with the 'what do you see' approach and then works round to a given meaning.

Then flip to the 'twenty questions' approach which interrogates the class till they come up with the 'right' answer.

It ends up with 'Here we have…' 'This is…' and 'As you can see…' with little alternative but to accept the given meaning irrespective of prior meanings.

Van Matre's analysis of deriving meaning from outdoor experience involves several aspects:

- Touching

- Tasting

- Hearing

- Smelling

- Seeing – and, here, interestingly he asks, how does what you are viewing make you feel? He emphasizes that seeing is not receiving. It is 'gathering light images and *interpreting them on the basis of past receptions*' (my italics) and realizes that 'to change our perception we must change our past'.

- Feeling

- Focusing – asks us to focus on the usual

- Framing – enclosing scenes we want to examine using fingers/hands. 'Establishing boundaries . . . narrows perception and sharpens patterns. . . .we see best when we don't try . . . we can gain impressions *undimmed by our recollections or verbalisations*' (my italics)

- Grouping – find the major elements in the scene

- Expanding – make a conscious effort to avoid looking at familiar objects because '*our eyes focus unconsciously but persistently on objects which we already know*' (my italics)

- Filling – look at spaces between solid objects

- Surveying – 'the familiar is unfamiliar from a different vantage point'.

- Observing – relax and let the natural world engulf you

- Orchestrating – fuse all the facets of awareness

- Scrutinizing – looking at small things

- Empathizing – identify with what you are observing

- Silencing – *forget about making sense of things*, let meanings seep away

- Waiting – let whatever occurs next happen.

Hug-a-tree hippie twaddle? I think not. This methodology is very perceptive, especially in terms of the things I have italicized above. Simply put, what he terms receptions, recollections, verbalizations and making sense of something refer to the constructs I discussed earlier and relate to the *tyranny of constructs*: familiarity directs our gaze and what we select as significant.

Now think of the field teaching approach: 'This is. . .' and 'Here we see . . .' That can be an enrichment, but it is also simultaneously an impoverishment through the fact that other meanings and interpretations are disallowed and not rewarded.

Now am I concluding that we should throw away didactic teaching and systematic investigation and follow the methodology of Van Matre? No, I am not. However, there is no reason why we should not follow it and, indeed, what Macnaghten and Urry referred to as 'auratic immersion' – sensing what is around you – is (a) a positive experience and (b) refers to the reasons why many of us might actually want to be in the field in the first place. What I am concluding is that we can learn a lot from this methodology and reinforce Feyerabend's (1999) conclusion that 'Variety disappears when subjected to scholarly analysis . . . Anyone who tries to make sense of a puzzling sequence of events . . . is forced to introduce ideas that are not in the events themselves but put them in perspective. There is no escape: *understanding a subject means transforming it . . .* – inserting it into a model or a theory or . . . account.' I am making three conclusions:

1. that *academic* interpretations, though enriching, tend to disempower other emotional significances

2. that within academia, given (or even derived, personal) meanings then disempower alternative (academic) interpretations

3. that in all cases it is through the emotions that we operate. Whether from 'auratic immersion', where the senses feed directly to the emotions, or the scientific meaning derived, say, from a graphical plot or something else that 'makes sense', it is the excitement, satisfaction or other 'emotional rush' which leads to a feeling of significance.

These give me both insights and concerns. First, field teaching should recognize that, as Davies (1999) recorded, we have to offer (and nurture) 'feelings of place (emotions, reactions, values) as well as knowledge about a place (information)'. This not least because the motivation for many physical geographers lies in feelings for a place or a kind of terrain, as I recorded at the outset. The excitement comes not only from 'making sense' of a landscape but also from 'being there'. This is important for personal meanings – the excitement of science and the excitement of experience.

Second, field teaching should empower alternative narratives and not just reward and reinforce singular, given meanings, negating all others. This is important for the subject. A given or derived meaning should never be seen as the *only* meaning, no matter how much sense it currently makes. Teaching of alternative, parallel and non-exclusive meanings is a methodology because it inculcates at an early age an important methodology for research – of keeping an open mind. Making sense of something is very seductive, but it is never the final word. We must beware of learning to make sense of something and then stopping thinking and looking.

The provisionality of conclusions and meanings is fundamental if the science is to progress to the development of new ideas. The worst kind of field teaching lays the foundation for the worst kind of research: such teaching just shows an expected pattern, worse still using the crutch of statistical testing, so it fossilizes rehearsed meanings and disempowers all others, focusing the attention in a way that makes all other possible meanings not legitimate and irrelevant. It is nice and rewarding when something makes sense, but since that 'sense' then directs our gaze, at all levels of investigation we should all be asking 'now, what am I not seeing?' The seductive compulsion to 'make things work' drives us on but makes us selective. While we are emotionally rewarded when 'it works', we should not feel the sense of failure when something 'does not work', but a sense of discovery. Be aware that your constructs, good, nice and necessary as they are, are directing your investigative gaze. Nature is there for its own sake and not for mine, though I make it what it is for me.

2.5 Summary

Constructs are means by which we give significance and meaning to the perceived world and:

- unavoidable, in order to have some interpretation and meaning
- good, nice, necessary, comforting
- ways of making sense of the world
- providers of motivations for action.

  We should be aware that they exist as they can:

- impoverish rather than enrich experience
- direct our attention at the expense of other items which might be observable
- focus and constrain our thoughts, excluding or marginalizing other ideas and meanings, and limit our possibilities of meanings and range of possible reactions and actions, e.g. in terms of framing a problem and finding a solution, setting up automatic responses which deny other possibilities.

Deciding what is an *appropriate* construct is always going to be contested. In physical geography, appropriateness might be judged in terms, verifiability and evidence. Constructs can be interpreted in terms of who is holding them ('they would say that, wouldn't they'), so one resolution is to group contested meanings (different schools of thought), but that still belies the variety of meanings.

Many central constructs are normative – widely held and accepted – but can still be contested. Any one construct about the way the environment works might be supported or refuted; what is important is to examine the **consequences of holding them**. Can, or indeed should, not a range of constructs be 'allowed' in a pluralistic endeavour which allows a range of different ideas to co-exist, rather than a convergence of thinking and the tyranny of paradigms: 'we all think . . .', 'we should all work on . . .'? Embodied in this chapter is the idea that it is the mavericks – the free thinkers – that lead us on. Approval by others could just be some kind of failure.

2.6 Consider the following exercise

Landform interpretation – *constructs of meaning for communication: description, classification, measurement, interpretation and testing*. Consideration of constructs is important as they act to control how we relate to, regard and treat the world. Many field groups are asked to measure something, often so a preconceived pattern emerges ('it worked'). I would like you think about approaches rather more deeply – is fieldwork a 'discovery'?

Recent writings have suggested that: 'information does not just exist "out there" waiting to be "collected" or "gathered" but is constructed, or created, in specific . . . contexts for particular purposes' (Mosse, 1994 quoted from Twyman et al., 1999). In addition, '(we) can transform our concept of the field from *a space in which we interpret and read the world*, others, objects, etc., to *a space through which the world is produced*. In other words, the field is not a point of access or gateway through which we read and represent the "other". Rather it is the fundamental process through which our own geographic world (as a meaningful narrative) is constituted' (Rose, 1999).

Description

How do you best describe this area so that someone else may visualize it and compare it with sites elsewhere?

- Descriptive phrases

- Different things at different scales

- Colour, shape, texture, etc.

- Photography/drawing

- Descriptive writing/poetry

- Classification of types (does classification aid communication or restrict it?).

Measurement and sampling

How does the form vary and thus what will be an appropriate scale at which to measure laterally, vertically?

• Continuous measurements or point measurements – if so, what interval?

• A A B B B C D D E E E F G G G G – actual

• A   B   B   D   E   E   G   G   samples:

• A   B   D   E   G   G

• A   B   E   G

• A   C   E   G

• A   D   G

• Quadrats/area.

Interpretation

Describe then interpret? What is a 'cause'?

• Use prior interpretation, describe with this in mind?

• What processes are involved and how do you tell?

• What is the relationship between process and form?

• Is what you show influenced by method?

• So how to choose the best methods?

• Look at the landforms. How do they vary?

• How do you describe to test theories of origin. What might the origins be?

• What might the causal factors be?

• How best to describe/measure causal factors?

• How do you measure to test? What do you measure and why? How do you measure?

• Is your selection designed to 'show' your theory or is it a fair test of it – and what is a fair test?

Reflections

In *Conquest of Abundance: A Tale of Abstraction Versus the Richness of Being* (1999), Feyerabend writes that 'Variety disappears when subjected to scholarly analysis. This is not the fault of scholars. Anyone who tries to make sense of a puzzling sequence of events, his or her own actions included, is forced to introduce ideas that are not in the events themselves, but put them in perspective. There is no escape; *understanding a subject means transforming it,* lifting it out . . . and inserting it into a model or theory.'

Pimm also writes: 'Theories tell us where to look and when we readily find what we are looking for, we gain confidence in both the theories ... and the data'. (Pimm 1991).

Have you been open minded? Have you found a pattern which relates to general principles or is your observation a unique case? Are you imposing a logic/order or making a discovery? Does fitting what you have found into a model preclude new ways of thinking? Has it 'worked' when you get a pattern or have we forced a construct upon it? Look at ALL the data: if what you find does not fit a model, do you reject what does not fit or revise the model?

Consider the parallels between tourism and research: should tourists actually read the guidebook first?

- Dissatisfaction; doubt experience?
- Don't doubt experience; reject observation/complain/criticise; revize model?
- What do you think?
- Have you taken photographs of scenery? If so, what of and why did you select it?
- How does knowledge of landscape origin contribute to the experience of landscape.? Does geomorphological/geological/biogeographical knowledge enrich the experience?

I have run this exercise on a rocky shore for some years now and received graphs of transects showing the relationships between morphology, colour zones, moisture retention and bioerosive algae. Moreover, student poetic writings and sketches have also evoked a sense of place and what it meant to them – feelings that are otherwise edited out, a theme I shall return to at the end of the last chapter.

References

Bishop, P. 1992: 'Rhetoric, memory and power: depth psychology and postmodern geography'. *Environment and Planning D: Society and Space*, 10, 5–22.

Brody, H. 2001: *The Other Side of Eden: Hunter-gatherers, Farmers and the Shaping of the World*. Faber and Faber, London.

Davies, A. 1999: 'Report of Discussion, Environmental Education and Citizenship Conference'. *Ethics, Place and Environment*, 2(1), 82–87.

Feyerabend, P. 1999: *Conquest of Abundance: A Tale of Abstraction Versus the Richness of Being*. University of Chicago Press, Chicago.

Kearns, G. 1998: 'The virtuous circle of facts and values in the New Western History'. *Annals of the Association of American Geographers*, 88, 3, 377– 409.

Macnaghten, P. & Urry, J. 1998: *Contested Natures*, Sage, London.

Mosse, D. 1994: 'Authority, gender and knowledge: Theoretical Reflections on the Practice of PRA'. *Development and Change*, 25, 497–526.

Pimm, S.L. 1991: *The Balance of Nature? Ecological Issues in the Conservation of Species and Communities*. Chicago University Press, Chicago.

Rose 1999: IBG Conference Abstracts.

Sack, R.D. 1990: 'The realm of meaning: the inadequacy of human-nature theory and the view of mass consumption'. Ch. 40 in Turner, B.L., Clark, W.C., Kates, R.W.,

Richards, J.F., Mathews, J.T. and Meyer, W.B., 1990: *The Earth as Transformed by Human Action: Global and Regional Changes in the Biosphere over the Past 300 Years.* Cambridge University Press, Cambridge, 659–671.

Trudgill, P. 1975: *Accent, Dialect and the School.* Arnold, London.

Twyman, C., Morrison, J. & Sporton, D. 1999: 'The final fifth: autobiography, reflexivity and interpretation in cross-cultural research'. *Area*, 31 (4), 313–325.

Van Matre, S. 1974: *Acclimatizing: A Personal and Reflective Approach to a Natural Relationship.* American Camping Association, Martinsville, Indiana.

Wischmeier, W.H. & Smith, D.D. 1978: *Predicting Rainfall Erosion Losses.* US Department of Agriculture, Agriculture Research Handbook, 537, 62.

PART II

PERSONAL MEANINGS

3
Realms of gold, wild surmise and wondering about physical geography

Tim Burt

Much have I travell'd in the realms of gold,
And many goodly states and kingdoms seen;
Round many western islands have I been
Which bards in fealty to Apollo hold.
Oft of one wide expanse had I been told
That deep-brow'd Homer ruled as his demesne;
Yet did I never breathe its pure serene
Till I heard Chapman speak out loud and bold:

John Keats: On first looking into Chapman's Homer

3.1 Introduction

First, please excuse a little personal history. Like Dick Chorley and Peter Haggett, I grew up in Somerset. I mention their names, not because I regard myself as being in the same league, but because background and upbringing are bound to be influential on one's later career. For reasons I explore below, Somerset was for me a rich place to grow up as a geographer. Later, I discovered that Chorley and Haggett had provided the foundation for my professional training, and indeed for what followed.

First, Somerset. My old geography teacher, Jim Hanwell, once asked me: 'Where did it first happen for you?' This was no prying into personal matters, but a genuine question about whether there had been some point on the road to Damascus where I suddenly realized I wanted to become a geographer. Unlike Jim, who cites Cheddar Gorge as his place in question, I cannot remember one particular place, although the cliffs at Charmouth loom large in my memory. Rather, I can think of two episodes, both redolent of space rather than place, which fired my interest. First, I remember being left in the car while my brother went off somewhere, leaving me with nothing to read but a copy of Bruce Sparks' *Geomorphology*, which must have been rather new then (my brother was then in the sixth form studying with the same aforementioned Jim Hanwell; I must have been about 10 at the time). I can still remember the fascination of reading about the Davisian Cycle for the first time. Sad, maybe, but true! Then, having ended up at the same grammar school, I recall, early on, being sent out to measure the width of the main road outside the school gate in order to compare with the red line printed on the Ordnance Survey map (an old 'One Inch' version, no doubt). Health and Safety concerns would preclude such activity these days, and clearly Jim should have been severely reprimanded for placing me in such peril but, nevertheless, I was hooked and, whether he knew it or not, Jim was destined to be stuck with me until the end of my school career and beyond. His influence was – and remains – very great, and I remain very grateful to him. Clearly, for some of us, charismatic school teachers have a lot to answer for! As for place, Somerset certainly played its part, imbuing me with a deep love of landscape, be it Mendip periclines, Jurassic escarpments, the Vallis Vale unconformity (which links those two), or the Levels (always rather foreign to a hill dweller like me).

Next, Chorley and Haggett. When I arrived at Cambridge, the white heat of the quantitative revolution in geography had cooled a little, but *Models in Geography* was still pretty new and the words 'model' and 'spatial' were everywhere. Whilst I had been made aware by Jim Hanwell that Geography was on the change, nevertheless the constraints of syllabus meant that my school studies had contained a good deal of the old geography, notably descriptive regional studies. It was clear from the outset at Cambridge that something very dramatic had happened and, though I may have been a few years too late to view the most heady days of the revolution, these were still exciting and unsettling times. I well remember my first attempts to struggle with *Models* in the autumn of 1970 and to make sense of all this new stuff. Dick Chorley and Barbara Kennedy (two more charismatic teachers) introduced us to 'Leopold, Wolman and Miller' and their own book, *Physical Geography: A Systems Approach,* was soon to appear. David Harvey's *Explanation in Geography* was hot off the press too, and Carson and Kirkby's *Hillslope Form and Process* came along in my third undergraduate year. As you can tell, we read a lot at Cambridge, which is no bad thing for anyone (c.f. Keats' comment on Chapman). A little later I went to Bristol to study for a Ph.D. There I met Peter Haggett and was lucky enough to experience the other end of the Cambridge–Bristol axis.

So, why this bout of personal self-indulgence? Peter Gould (1985, p320) encourages us to recognize the particular historical situation in which we find ourselves, 'to acknowledge the trap of one's own time and place and circumstance'. Well, this is my time and place and circumstance: in terms of 'normal science' (Kuhn, 1962), you will now be able to appreciate where my own brand of quantitative physical geography has come from. Why I have studied what I have in the way I have. Gould continues, noting that 'the vanguard of one era becomes the rearguard of the next ... ideas begin as liberating forces and evolve in the course of time into suffocating straitjackets'. Well, I was certainly around when the quantitative revolution was still quite new and it has coloured my research career without doubt. Not quite in the vanguard maybe, but close enough to be caught in some of the crossfire! And as for liberation and suffocation, more of them anon.

3.2 What sort of scientist am I?

> At the beginning of any inquiry, there is always a choice to be made, the choice of what shall be the things and relations forming the subject of the inquiry itself ... Whatever the focus of inquiry, choices have to be made that some things are relevant, while others are not.
>
> Peter Gould, 1985, p324

Dick Chorley often mentioned in his lectures that two of the most influential papers in geographical literature had in common that they were both published posthumously. Robert Elmer Horton's 1945 paper was a landmark in fluvial geomorphology, laying out a theory of hillslope erosion and encouraging the quantitative study of process and form. Fred K. Schaeffer's 1953 publication advocated transforming geography from an idiographic discipline to a nomothetic science. Recondite vocabulary aside, this means moving from a descriptive endeavour with few organizing principles to one with predictive general theories. The opening chapter of *Models* says much the same and, for me, these ideas were cemented in David Harvey's *Explanation in Geography*.

The aim of scientific explanation is to establish a general statement that covers the behaviour of the objects or events with which the science is concerned, thereby allowing a connection to be made between separate known events and reliable predictions to be given about events as yet unknown. David Harvey (1969) showed that there had been a revolution in the way geographers arrive at an explanation. In the inductive or 'Baconian' route (Figure 3.1), explanations are obtained through experiment and observation; the explanations depend on the facts available and the way in which they are organized. Prior to the quantitative revolution, geography was dominated by regional description, collecting and indexing facts about particular parts of the world with very little attempt to extract general principles. In physical geography there was at least a general model, the geographical cycle of William Morris Davis. However, this served in practice only as a basis for classifying landforms according to their stage of development, and led inexorably to unique accounts of the system in question (i.e. the denudation chronology of a particular site). What cannot be done with such conclusions is to apply them safely to other sites or to provide predictions of future events.

The weakness of the inductive route to explanation is that general statements cannot be based upon specific instances. The deductive route to explanation (Figure 3.1) depends upon a clear distinction between the origin and testing of an idea. Faced with a problem, the scientist attempts to solve it by putting forward a trial solution or hypothesis (Haines-Young and Petch, 1986). If the trial solution is true, certain hypotheses must follow logically. These can then be tested in an attempt to prove the trial solution false. The best established ideas are those that have withstood a gruelling procedure of testing. The important point is that the deductive route to explanation – in other words, scientific method as we generally recognize it today – must, from the

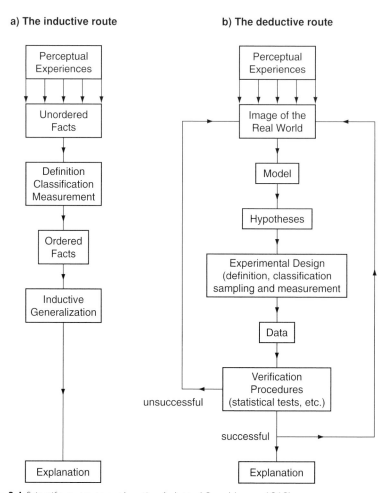

a) The inductive route

Perceptual Experiences

Unordered Facts

Definition Classification Measurement

Ordered Facts

Inductive Generalization

Explanation

b) The deductive route

Perceptual Experiences

Image of the Real World

Model

Hypotheses

Experimental Design (definition, classification sampling and measurement

Data

Verification Procedures (statistical tests, etc.)

unsuccessful

successful

Explanation

Figure 3.1 *Scientific routes to explanation (adapted from Harvey, 1969).*

outset, be seen as a complete package, not a series of disparate parts. Investigations are hypothesis-driven, with data collection designed to address the problem at hand. Although controlled laboratory experiments do not transfer easily to the natural environment, nevertheless physical geographers have devised a number of approaches to data collection, including plot studies and paired catchments. So-called 'experiments' are, strictly speaking, not always precisely so; Church (1984) distinguishes between exploratory studies, where initial testing of ideas takes place, and confirmatory studies, where a tighter experimental design is required.

The adoption of the deductive route to scientific explanation can be seen as both a cause and effect of the growing interest in systems analysis that affected physical geography from the early 1960s (e.g. Chorley, 1962). It gave a framework – some might even argue, a paradigm (Kuhn, 1962) – within which individual case studies could be carried out. Chorley and Kennedy's *Physical Geography: A Systems Approach* appeared not long after *Explanation in Geography* and provided the organizational basis for a wide variety of studies; a number of other similar texts have appeared since. Systems analysis can be summarized as follows:

- to understand how particular systems function and evolve
- to make measurements of the state of the system and its process activity
- to use statistical methods as a means of testing hypotheses and
- to control and manage the physical environment.

One important outcome of the growing preoccupation with process – something that had been totally lacking in the Davisian approach – was a reduction in the scale of interest. As will be seen later, there have recently been signs that the scale of attention is enlarging again, as people begin to synthesize their results and seek to apply their theories to larger landscape units.

One problem remains, signposted in the Gould quotation at the start of this section: what to study and what to ignore? As noted above, the weakness of the inductive method is that general theories cannot be based on particular instances. And yet it is from such observations that we gain our experience of the real world. What is often lacking in any discussion of the deductive route to explanation is any consideration of where ideas come from in the first place. Preliminary experience remains vital – it is from such observations that we gain our experience of the real world. What we must do is to use these experiences as the basis for general statements that will cover all objects or events. An isolated fact is of little interest (see figure 3.3 below), but a set of related facts, when set against our previous experience, can become an important ingredient of science if it suggests some sort of relationship or forms the basis for a more general statement. Sometimes, of course, we can think of several possible explanations for an event; such competing ideas (Chamberlain's 'multiple working hypotheses') must then be tested, discarding those that do not stand up to scrutiny and retaining the last remaining hypothesis for further testing and refinement. So, induction remains vital to us, not as the overarching structure for our scientific investigations, but as the preliminary activity that fires our interest and focuses our research. New ideas need to come from somewhere!

Of course, we all work within a particular context, socio-political as well as scientific. My own brand of physical geography, most of it concerned with catchment hydrology and water quality, clearly has an applied slant, aimed ultimately at improving the management of the physical environment. At the same time, my research is rooted in deductive science and systems analysis, reliant on an understanding of process mechanics and informed by quantitative data. I have no doubt about the fact that my science must contribute to society, but it is science nevertheless. The broad context of my work – 'normal science' in Kuhnian terms – has its foundation in the functional and realist approaches to geomorphology that developed in the 1960s as part of the post-Davisian quantitative revolution (Chorley, 1978). I am a hillslope geomorphologist at heart, fundamentally interested in hillslope hydrology and fond of small catchments. However, such research can and does form the platform for the study of larger river basins, and so is not stuck in the cul-de-sac of reductionism (a jibe often thrown at me by my good friend, Andrew Goudie!). But we all need new ideas from time to time – so where do they come from?

3.3 Where do our ideas come from?

I doubt there is any magic formula for finding new ideas, simply a curious eye. Therefore, some examples will suffice.

3.3.1 Seeing is believing – or believing is seeing?

I spent a weekend recently staying with Professor Denys Brunsden on the west Dorset coast, looking at evidence of landslide activity from the previous, remarkably wet 2000/01 winter. Denys, who knows more about the landslides and mudflows of this area than anyone else, had been surprised by one particular process, something he had long known about but thought unimportant in the grand scheme of things. The great 1839 landslip to the west of Lyme Regis was described in detail by some of the early 'greats' in geology and geomorphology, including Buckland and Conybeare. They had described how liquid sand spurted out from the base of the Greensand (which forms the uppermost part of the cliff section), undermining the overlying rock and immediately causing the upper cliff section to fail. It was only when he saw this for himself last winter that Denys realized the significance of those early observations. Previously, he had interpreted these landslides in a classic manner, invoking standard theory. However, they are not simple rotational slides: the cavitation process that accompanies the liquefaction of the very wet sand provides a new twist to the model. The evidence had been there all along in the literature, but it can sometimes be all too easy to dismiss an idea in the face of what seems like more compelling evidence. The problem with models, then, is that they colour our view – we see what we believe we see.

A second example comes from my own experience of teaching hillslope geomorphology in south Devon. Having spent a good deal of time looking at hydrological processes, I then like to ask the students: 'How old is this valley?' This is a useful way of broadening the perspective, in space and, especially, in time. It allows me to draw upon Mike Kirkby's computer simulations of hillslope form and evolution (Kirkby, 1984) and even to retreat to pre-quantitative revolution days with mention of erosion surfaces (the late Tertiary erosion surface of south Devon is unmistakable and unarguable, cutting as it does across folded Devonian strata). On one occasion we revisited this theme whilst discussing the evolution of coastal landforms in the same region. Figure 3.2 shows a generalized profile: tor-like pinnacles; a rectilinear slope

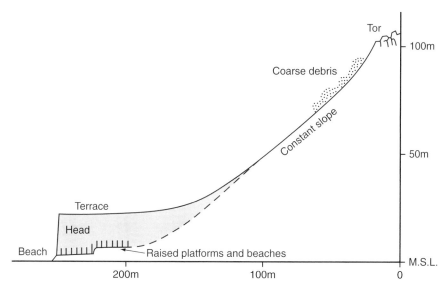

Figure 3.2 *A generalized slope profile of the south Devon (UK) coast.*

mantled in coarse debris; raised shore platforms often covered by solifluction deposits; plus contemporary valley and beach landforms. When I questioned the idea that the tors might be an old cliff line, one bright spark (a young David Higgitt, now one of my colleagues at Durham) swiftly replied that they were surely very different to any slopes we had previously looked at. He was right, of course – I was looking only for signs of fluvial erosion, and I could not see the rather different evidence of relict marine action right in front of my nose!

There is no substitute for fieldwork, in my opinion. I continually repeat this mantra, mainly on the back page of *Geography Review*, for anyone who will listen. The old adage of getting mud on your boots – or, for human geographers, chewing gum on your trainers – remains as apposite as ever. One more example: I remember being summoned into the field by a graduate student at Oxford, Mike Slattery, who wanted to show me the first evidence of rill erosion at his field site. I showed enthusiasm, of course, but in truth, his few modest rills by the roadside were unimpressively small – size does matter! I suggested we looked over the next rise to see what had gone on in an adjoining dry valley. We were not disappointed: a large thalweg rill had eroded along the valley floor. Mike may have been frustrated at not having spotted it himself, but we had both learned something new about our field site and added a new element to the study (see Slattery et al., 1994). Thalweg rills were by no means new to science, but preoccupation with 'Hortonian' processes meant that they were unexpected. As with Denys Brunsden's landslide, we needed to build a more elaborate model of soil erosion in our catchment – seeing was believing!

3.3.2 So what?

It may seem odd to turn from the natural landscape – the raw material of our subject – to data as a source of ideas. True, this is one step removed from reality, but in my experience data archives can very often prove a fruitful source of new ideas (see Burt, 1994). There is a huge amount of data available, much of it gathering dust on the shelf. Some of this has been collected by keen amateurs (rainfall data is a good example) whilst other material results from statutory monitoring (e.g. to check drinking water quality); in both cases the quality of the information is usually high. Moreover, these time series are invariably long, giving the opportunity to detect various patterns of change.

Magnuson et al. (1991) show the value of long-term data in their analysis of ice cover on Lake Mendota. Ice cover in one winter is relatively uninteresting, uninterpretable and, for that matter, soon forgotten. Yet, when the record is extended to 10 years, 50 years, or the entire 132-year record, one winter can be put into context and can be better understood. Importantly, new ideas emerge, in their case, the recognition that El Niño years all tend to have shorter durations of ice cover; and, more recently, a global warming signal has also emerged. Using the same approach, Figure 3.3 shows 12-month rainfall totals at Durham for all months since December 1852, plus a six-year running mean. (We are well used to totals for calendar years, ending in December; this graph's 12-month total for every month since the record began is in effect, a 12-month running mean.)

Take November 1989: the 12-month total from December 1988 to November 1989 is 379.5 mm. So what? It happens to be a great deal below the long-term average (662 mm), but if we only have one month's data, we do not appreciate even that. But even knowing the annual mean, we cannot tell how rare such a drought episode might

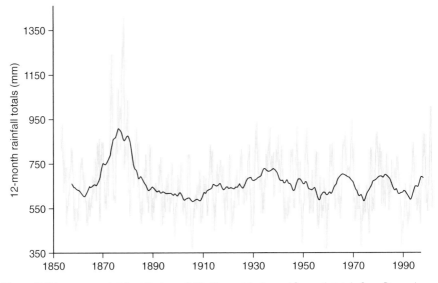

Figure 3.3 *Long-term rainfall at Durham (UK). The graph shows 12-month totals from December 1852, together with a 10-year running mean.*

be. In fact, the graph shows that such droughts have happened before, but not very often. If we now study the period from 1989 to the end of the record, what now? There are wet and dry phases but seemingly an overall upward trend. Is this significant? Can we be tempted to postulate some connection between increasing rainfall and global warming? Taking the whole record since 1852, we can see that the 1990s seem not much different to the rest of the record. However, we can also see that the very wet 2000/01 winter has the highest totals since the 1870s, extreme by any standard – as were the floods that followed. We can be confident of the accuracy of the recent numbers, but what of the early data, particularly the unusual 1870s. Data from Edinburgh and a composite record for North East England show similar, very high totals in the 1870s, seeming to confirm the Durham figures. So, what was happening in the 1870s? What patterns of global circulation prevailed in the North Atlantic at that time? How does that period compare with the 1990s? There are no answers as yet, but one salient feature of the 1990s (not shown on Figure 3.3 and not evident in the 1870s) is an increasing seasonality, a tendency for wetter winters and drier summers. This knowledge has proven helpful in explaining water quality during and after droughts (see Figure 3.4 and below). The important conclusion is that new questions have emerged from the data for us to ponder, taking our science forward in new and sometimes unexpected directions. Given the noisy record of environmental change and the subtle signals involved, it takes time to detect these changes, but it's well worth the wait!

The long nitrate record for streams draining into Slapton Ley, a eutrophic freshwater lake in South Devon, has proven equally interesting (Figure 3.4). Water samples have been taken each week since 1970 in an attempt to produce a nutrient budget for the

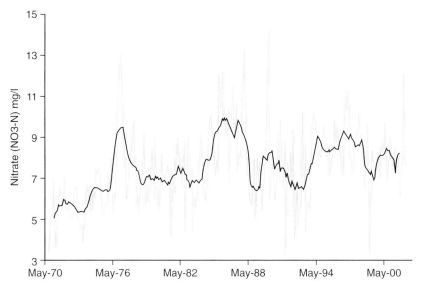

Figure 3.4 *The long-term nitrate record for the Slapton Wood catchment, south Devon (UK). Average monthly values are shown together with a 12-month running mean.*

lake, to understand its nutrient dynamics and to monitor unwelcome changes in lake water quality. The emerging record has revealed evidence of important patterns to be explained; field experiments have been necessary to achieve this, demonstrating, for example, links between hillslope hydrology and nitrate leaching (Burt and Arkell, 1987). There is clear evidence in the record of an upward trend in nitrate, probably related to changes in land use and farming practice. Application of nutrient export coefficient models supports this hypothesis (Johnes and Burt, 1993). The record has also revealed evidence of a 'memory effect' within the system, with large amounts of nitrate leached in wet winters following dry summers, and vice versa. This has focused attention on soil nutrient dynamics, searching for process mechanisms to explain the observations. In industry, monitoring implies control – if a system is seen to be moving away from its optimal condition, something must be adjusted to bring it back into line. This is possible in drainage basin systems too, although perhaps less obviously. The link between land use and nitrate leaching can be exploited at two scales (Burt and Haycock, 1993): either widespread land use change can be effected across the entire basin or localized buffer zones can be established to protect the water course. In either case, we have only just begun to appreciate that our small-catchment studies on land use and nitrate leaching have relevance at the larger scale. Research at Slapton illustrates the whole spectrum of scientific method, therefore – from inductive investigation of long-term data, through hypothesis testing and field experiments, to using predictive models to manage environmental systems. It may be noted that the scale of investigation varies accordingly: we start with a knowledge of nitrate export from the basin (but no knowledge of internal dynamics), then reduce the scale of our attention to process mechanics at the hillslope scale, and finally move back to a distributed view of the river basin as we seek to model and control nitrate losses.

3.3.3 Failed experiments

As Figure 3.1 shows, if our hypothesis testing is unsuccessful, we must alter our image of the real world and start again. In practice, this usually means some slight adjustment to our ideas rather than discarding the whole thing and starting again. In this case, new ideas come from our failures, not our successes. One somewhat eccentric example will suffice. For my undergraduate dissertation I studied terracettes, parallel ridges that run around steep grassy hillsides. Conventional wisdom (geography textbooks) holds that terracettes are small-scale mass movement features. However, my field studies revealed no evidence of shear planes beneath the terracettes. There were other problems with the mini-landslide theory too. Why are terracettes so regularly spaced? Why do they converge on gateways? Why are there none in woodland. Alternative conventional wisdom (what farmers tell you) holds that terracettes are animal tracks. Sheep and cattle cannot walk straight up steep hills, so tend to move more or less horizontally. Following forest clearance, the original slopes may well have contained mini-landslides, but these have been evened out by the patter of tiny feet over many years. One clinching argument for me is a statistical one: animals can reach further upslope on steeper slopes, so terracettes are more widely spaced. Of course, my hypothesis remains untested. I am still searching for grassy but ungrazed slopes with regular terracettes on them. Railway and motorway cuttings lack them – check for yourself! Perhaps it needs a proper experiment: clear a steep slope of trees, turn in the animals, and see what happens. Trouble is, in the current climate, I doubt I could find anyone to fund my research idea – but that's another story.

3.4 Upscaling our science

These days, physical geography is firmly ensconced as a deductive science. Not one of the 'hard' sciences, of course, whatever that means, but more than complex enough for most of us (note here the probably apocryphal story that Einstein took up physics because he thought that geography would be too difficult). One need only think about the words 'risk' and 'uncertainty' to show how difficult our science is. The debate over global warming demonstrates an interesting progression from a sound theory (increased levels of various greenhouse gases), through fairly good model predictions at the global scale, to much uncertainty about detail at the regional scale. How do we marry scientific uncertainty with the need to provide politicians with firm answers. How do we educate the public to extend their knowledge of uncertain weather forecasts to longer timescales? What is the risk of global warming? What is the added risk if we try to do nothing about it? Peter Gould (1985, p326) reminds us that we cannot detach our science from the wider concerns of society, quite the converse. I am sure that risk and uncertainty will be major themes for us in the near future.

Much of our science is, for better and worse, reductionist in scale, grappling with the details of process and response on a small hillslope or a single riffle-and-pool sequence. Even so, there are some signs of larger-scale work beginning to emerge: studies of global climate change and teleconnections (Phillips et al., 1999), geomorphological studies that model entire landform sequences (Kirkby, 1984) or infer their evolution over long periods of time (Seidl et al., 1997), Quaternary studies that correlate sea-level change over long lengths of coastline (Shennan and Woodworth, 1992) and hydrological models that build from knowledge of process hydrology to encompass entire river basins (Whitehead et al., 1998). Maybe our perspective is beginning to broaden again, founded not on sterile debate but on a well-tested understanding of system dynamics. It may be that the grander scale may prove more relevant, as well as providing more awe and wonder about the world around us. As far as I can tell, the systems paradigm remains invigorating not suffocating, but the need for new ideas remains, and for that a close appreciation of the geographical world is vital. These days, computers do so much for us (including the preparation of this chapter, of course), but there are limits and no virtual reality (not even long data sets) can replace real fieldwork and deep curiosity about our world. I guess this is my main message. I have been lucky: brought up in stimulating rural landscape and then given the opportunity to travel, as Gould says, with unbridled hedonistic delight!

3.5 Acknowledgements

I take this opportunity to acknowledge my debt to various teachers and tutors, including Jim Hanwell, Dick Chorley, Barbara Kennedy, David Stoddart and Malcolm Anderson. Peter Haggett has never taught me *sensu stricto*, but he has always given strong encouragement – I showed him deer on the Quantocks in return! I thank Denys Brunsden and Mike Slattery for permission to tell stories at their expense and for stimulating times together in the field. Years ago I used to read Arthur Ransome; the first quotation in *Swallows and Amazons* is the last one included here, the final lines of the sonnet quoted at the start as it happens. A little 'wild surmise' does us all good from time to time.

> *Or like stout Cortez when with eagle eyes*
> *He stared at the Pacific – and all his men*
> *Look'd at each other with a wild surmise*
> *Silent, upon a peak in Darien.*

References

Burt, T.P. 1994: 'Long-term study of the natural environment: perceptive science or mindless monitoring?' *Progress in Physical Geography* 18, 475–496.

Burt, T.P. and Arkell, B.P. 1987: 'Temporal and spatial patterns of nitrate losses from an agricultural catchment'. *Soil Use and Management*, 3, 138–143.

Burt, T.P. and Haycock, N.E. 1993: 'Controlling losses of nitrate by changing land use'. *Nitrate: Processes, Patterns and Management*, T.P. Burt, A.L. Heathwaite and S.T. Trudgill (eds.), John Wiley, Chichester 341–367.

Carson, M.A. and Kirkby, M.J. 1972: *Hillslope Form and Process*. Cambridge University Press, Cambridge.

Chorley, R.J. 1962: 'Geomorphology and general systems theory'. *United States Geological Survey Professional Paper*, 500-B.

Chorley, R.J. 1978: 'Bases for theory in geomorphology'. *Geomorphology: Present Problems and Future Prospects*, C. Embleton, D. Brunsden and D.K.C. Jones, (eds.) Oxford University Press, Oxford, 1–13.

Chorley, R.J. and Kennedy, B.A. 1971: *Physical Geography: A Systems Approach*. Prentice-Hall, London.

Church, M.A. 1984: 'On experimental method in geomorphology'. *Catchment Experiments in Fluvial Geomorphology*, T.P. Burt and D.E. Walling, (eds.) GeoBooks, Norwich, 563–80.

Gould, P. 1985: *The Geographer at Work*. Routledge & Kegan Paul, London.

Haines-Young, R.H. and Petch, J.R. 1986: *Physical Geography: Its Nature and Methods*. Harper and Row, London.

Harvey, D. 1969: *Explanation in Geography*. Edward Arnold, London.

Horton, R.E. 1945: 'Erosional development of streams and their drainage basins: hydrophysical approach to quantitative morphology'. *Bulletin of the Geological Society of America*, 56, 275–370.

Johnes, P.J. and Burt, T.P. 1993: 'Nitrate in surface waters'. *Nitrate: Processes, Patterns and Management*, T.P. Burt, A.L. Heathwaite and S.T. Trudgill (eds.), John Wiley, Chichester 269–317.

Kirkby, M.J. 1984: 'Modelling cliff development in South Wales: Savigear re-viewed'. *Zeitschrift fur Geomorphologie* 28, 405–426.

Kuhn, T.S. 1962: *The Structure of Scientific Revolutions*. University of Chicago Press, Chicago.

Leopold, L.B., Wolman, M.G. and Miller, J.P. 1964: *Fluvial Processes in Geomorphology*. Freeman, San Francisco.

Magnuson, J.J., Kratz, T.K., Frost, T.M., Bowser, C.J., Benson, B.J. and Nero, R. 1991: 'Expanding the temporal and spatial scales of ecological research and comparison of divergent ecosystems: roles for LTER in the United States'. *Long-term Ecological Research, an International Perspective*, P.G. Risser (ed.), SCOPE 47, John Wiley, Chichester, 45–70.

Phillips, J., Rajagopalan, B., Cane, M. and Rosenzweig, C. 1999: 'The role of ENSO in determining climate and maize yield variability in the U.S. Cornbelt'. *International Journal of Climatology* 19, 877–888.

Schaeffer, F.K. 1953: 'Exceptionalism in geography: a methodological examination'. *Annals, Association of American Geographers* 46, 226–49.

Seidl, M.A., Finkel, R.C., Caffee, M.W., Hudsin, G.B. and Dietrich, W.E. 1997: 'Cosmogenic isotope analyses applied to river longitudinal profile evolution: problems and interpretations'. *Earth Surface Processes and Landforms* 22, 195–209.

Shennan, I. and Woodworth, P.L. 1992: 'A comparison of late Holocene and twentieth-century sea-level trends from the UK and North Sea region'. *Geophysical Journal International* 109, 96–105.

Slattery, M.C., Burt, T.P. and Boardman, J. 1994: 'Rill erosion along the thalweg of a hillslope hollow: a case study from the Cotswold Hills, central England'. *Earth Surface Processes and Landforms*, 19, 377–85.

Whitehead, P.C., Wilson, E.J. and Butterfield, D. 1998: 'A semi-distributed Integrated Nitrogen model for multiple source assessment in Catchments (INCA): Part 1 – model structure and process equations'. *Science of the Total Environment* 210/211, 547–558.

4

Goodbye to Geographical Reality: a retrospect on the New Geography

Tim Bayliss-Smith

On writing Goodbye

This chapter is presented somewhat in the spirit of *Goodbye to All That*, the autobiography which Robert Graves wrote at the age of 33 and which he intended, as he said in the preface, to be 'a formal goodbye to you and to you and to you and to me and to all that'. Most of Graves' book is about his experiences in the Great War, and he wrote in order to achieve 'forgetfulness, because once all this has been settled in my mind and written down it need never be thought about again' (R.R. Graves, 1929, 13). I do not need to forget my own past, but I do want to reconsider it. It is now clear to me that the kind of geography I pursued as a graduate student was misconceived, even though at the time it seemed exciting, cutting-edge, and even 'relevant'. I now believe that I was seeking to define 'geographical reality' in the wrong way – so if not quite *Goodbye to All That*, for me it is certainly 'Goodbye to Geographical Reality'.

There are other ways in which Graves' book provides for me a useful starting point. Although he intended to write about his early experiences in a wholly frank and honest way, in the treatment of historical facts he is not always accurate. Partly this reflects Graves' view that to an artist the literal truth is relatively unimportant, but there were other reasons for his distortions, reasons that may be intrinsic to the writing of autobiography. By 1929 it was only possible for Robert Graves to write about his early life by means of a significant reconstruction of his past. Since the end of the Great War he had been tormented, not only by shell-shock but also by divided loyalties and irreconcilable beliefs. Events in the late 1920s culminated in a personal crisis, out of which there emerged what his biographer describes as 'a decisive new alignment of ideas, thoughts and feelings' (R.P. Graves, 1990, 102). This new world-view provided an opportunity for Graves to look back on his early life with a cool and cynical detachment, and from a new-found conviction that traditional values were meaningless because the 20th century was witnessing the end of history. There may be a parallel process of distortion in what I write in this chapter.

The writings of Robert Graves are not only an example of how one can achieve personal renewal, but they also provide us with an interesting metaphor. It is a metaphor about geographical reality that seems to have emerged out of Graves' wartime experiences, because for him geography was basically map reading. It was the means whereby spatial reality could be defined in a two-dimensional way. But the question that Graves wanted to ask was about people, not places – how can a person's 'reality' be defined? And what about oneself – how can the task of autobiography be carried out? For Graves the task was complicated by his growing conviction that some people have more depth of reality than others. He therefore makes a distinction between 'proper chaps', those people who have an existence that goes beyond their two-dimensional definition, and all the rest, the people who have only 'geographical reality'.

I believe that some of my own efforts to do geography have been based too much on seeing reality in terms of two-dimensional models, so I am attracted to this metaphor, even though I reject its élitist view of humanity.[1] Graves used his metaphor to explain the difference between two kinds of people, 'insiders' and 'outsiders'. The outsiders are the two-dimensional people, the people who can be described purely in terms of their 'formal geography'. Some of these people become prominent, sometimes they manage to 'put themselves on the contemporary map as geographical features', but ultimately they are unnecessary, they are 'the irrelevant people, the people with only geographical reality' (R.R. Graves 1929, 13–14).

The second type are those 'proper people', the insiders who recognize that our perception of reality has been transformed by the onset of modernity. Graves felt that modernity represented such a radical break with the past that historical time had, in effect, come to an end. Proper people know that:

> ... The bottom of things, after working looser and looser for centuries, has at last fallen out ... One unnecessary person too many was born ... The proper people were finally swamped. So it's impossible for a proper person to feel the world as a necessary world – an intelligible world in which there's any hope or fear for the future – a world worth bothering about.
>
> R.R. Graves, 1930, 293

This bleak vision of a material world that has become meaningless foreshadows some of the recent debate about globalization and the condition of post-modernity. It seems likely that Graves was drawn to this idea of insiders/outsiders through his relationship with the American poet, Laura Riding.[2]

Graves and Riding had begun collaborating together on poetry in 1926, a working partnership which later developed into an all-consuming love affair. Under Laura Riding's

---

1 Robert Graves too came to reject it. He expressed these views in the opening paragraphs of *Goodbye to All That* (1929) and also in his play, *But It Still Goes On* (1930), but he removed them from the revised edition of his book (1957) that people now read.

2 In later years, with the rise of fascism and the renewed threat of war, the idea expanded into a political manifesto which was circulated in 1938 by Riding, Graves and others. The manifesto was written after they had held meetings that were intended, as they put it, 'to decide on moral action to be taken by inside people: for outside disorders' (R.P. Graves, 1990, 285).

influence, Robert Graves began to reassess all the conventional values that had hitherto dominated his thinking, and as a result he entered into a period of profound personal crisis. The writing of *Goodbye to All That* was clearly a response to this crisis, but it was not done, he said, in order to 'put himself on the contemporary map' as a kind of geographical feature. Instead, he needed to write about his life in order to cast off all his previous historical existence. The task of autobiography was easy 'when loyalties have become negligible and friends … happen to be chaps for whom geography is also without significance'. But while seeking to go deeper into his own past, to investigate beyond the surface features of the historical landscape, he also considered that it was impossible to avoid altogether a conventional 'geography': 'And yet even proper chaps have their formal geography, however little it may mean to them. They have birth certificates, passports, relatives, earliest recollections…' (R.R. Graves, 1929, 13–14) Only by exploring this formal world could he re-define his attitude towards it and, in the process, recreate himself.

This chapter, then, is an attempt to examine the origins of my own 'formal geography', as a necessary prelude to some reconstructed view of geographical reality. Academics are not really encouraged to do this sort of reconstruction except in a silent way. At some stage almost all of us quietly jettison old research projects or feel embarrassed by what we have previously written. We delete publications from our reference list as we rewrite the CV. I would like to say goodbye to the past in a more explicit way, and for several reasons. I do find it interesting that the ways we have found to 'do geography' are perpetually being re-defined, and I think it might be worth exploring this process through my own ordinary, confused and probably typical career.

4.1 Geographical Reality: the view from Mallorca

It was in Mallorca in February 1994 that I first discussed with Linda McDowell my growing sense of 'geographical unreality'. We were visiting the island to make preparations for a student field course, and our discussions about how we could 'do geography' in Mallorca reached some sort of an impasse during our visit to Deiá. This was the village that provided a home for Robert Graves and Laura Riding from 1929 until the outbreak of the Spanish civil war in 1936. They lived there in a stone cottage, kept two fires blazing throughout the winter, cooked over a charcoal stove, grew vegetables, swam in the sea, ate olives, figs, oranges and fish. Above all, they worked. It was in the Casa Salerosa in Deiá that Graves corrected the proofs of *Goodbye to All That*. In the years that followed he found he was able to reconstruct his creative world in symbolic exile from his home, his children and his friends. He returned to Deiá after the Second World War and lived there for the rest of his life.

I visited Deiá with Linda McDowell during a day that had provided plentiful reminders of the changing face of Mallorcan geography. In Cala Millor in the early morning we had seen where the Pleistocene raised shorelines had been bulldozed to make way for the Cocacabana Cocktail Bar and the Polynesian Pub. Nearby was a prehistoric stone dwelling, a Bronze Age *talayot*, but the archaeological site had been virtually encircled by the glass and concrete of the Waikiki Hotel and was transformed into an adventure playground. Where in the world are we, I wondered? The same question had occurred to me the previous day at the airport when I had bumped my luggage cart against a German tourist. He and I looked at each other for a moment in linguistic confusion. 'Scusi', I muttered in bad Italian, making way for him. 'Merci', he replied in French. Linda had laughed and called it 'a typical case of time–space compression'. That day in Cala Millor I remembered these words as I posed for a photograph beside the life-sized statue of an islander, not a Bronze Age Mallorcan but the grass-skirted Polynesian who was guarding the pub. I had done my first geography fieldwork in the Pacific among Polynesian islanders, and it was disconcerting to encounter one in the Mediterranean. What we were experiencing in Mallorca was not so much time–space compression as time–space confusion, possibly even the end of geographical reality.

We needed some maps to guide us through this unfamiliar landscape, so later that morning we went to a government office in Palma to see what was available. The sheet that we wanted did not exist except in invisible form, as digital information in a memory bank, continuously updated but hidden from view inside a giant computer. However, at the flick of a switch a more familiar version of geographical reality began to emerge. As that map was printed out by the machine, line by line, Linda and I were both struck by the settlement pattern that it revealed. There was an extraordinarily regular spacing to the inland villages of Mallorca. It was very like the landscape of 'central places' that we had learnt so much about as students in the late 1960s, a theoretical landscape which to most geographers today is as remote and forgotten as a slide rule to an engineer. These days, Linda and I do very different kinds of geography, but we both still have the same instinctive reaction to patterns on maps. Probably no British geographer trained in the last 10 or 15 years would share our response. Today the 'proper' geographers (*sensu* Graves, 1930, 14) have discarded central place theory as misleading, ahistorical, economistic, something only talked about by 'all the irrelevant people, the people with only geographical reality'. Yet central place theory was an important part of the training of every geography student a generation ago. Had Linda and I visited Deiá as teenagers in the 1960s, and had we bumped into Robert Graves in the village bar, it would have been an aspect of geographical reality that we would have been pleased to explain to him.

In the afternoon we drove through the plain of central places towards the mountains, passing through an ancient cultural landscape of peasant agriculture. Around Valdemossa and Deiá the olive terraces were crumbling and overgrown, as the hard-won landesque capital of the past was being written off in the new political economy of the European Union. In the mountains at Son Moragues we visited a conservation area to have a look at ice storage pits, lime kilns and the sites of charcoal-burning in the oak forests. A heritage group, ICONA, has reconstructed the mountain hut where in winter the snow-gatherers used to live, making the ice that they would sell in Palma during the hot months of summer. According to the guidebook it was the work of men and donkeys.

On the February afternoon of our visit to Son Moragues, a new generation of Mallorcans was visiting the snow-gatherers' hut. About 20 youths were enjoying a noisy picnic in the sunshine, playing heavy-metal music on their stereos and scattering their plastic litter. It was quite a warm day but they were fashionably dressed as if for an Alpine resort: designer jeans, multi-coloured jackets, skiing gloves. Peering inside the dark interior of the snow-gatherers' hut, we disturbed a couple who were making love by the fireplace. The inept fire that they had thrown together was struggling to burn, a smouldering heap of picnic paper, old magazines and green branches torn from an oak tree in the forest.

This encounter seemed to symbolize all the symptoms of the globalization that has swept through Mallorca. Not only the tourists in the Polynesian Pub, but also the younger generation of locals, are now cut off from anything that has its meaning rooted in local production, local culture or local ecology. The entire landscape of Mallorca is becoming a neglected theme park. And what is striking is how quickly this has happened. It was probably the grandmothers of these teenagers that Robert Graves was writing about in 1936: 'Deyá is suffering from their being no demand from Barcelona for gloves, which are hand-made here by the women. But no real poverty anywhere: only anxiety.' (R.R. Graves, in O'Prey, 1982, 269–270).

Was it post-modern anxiety that made us try again to put some distance between ourselves and the real Mallorca? We turned our backs on the snow-gatherers' hut and climbed up to the mountain plateau in late afternoon sunshine, until we reached a point where we could look down on the village of Deiá far below. Ancient stone houses were clustered around the church, and the village itself was at the centre of a zone of agricultural terraces stretching from the oak forests of the mountains down to the shores of the Mediterranean. There seemed to be a self-evident logic in the land use pattern. It was possible to imagine, from this more remote viewpoint, that nothing fundamental had changed in this cultural landscape.

Our viewpoint that sunny evening was, in one sense, the same as that of Robert Graves who in June 1930 walked up from Deiá to exactly the same spot. In a letter he wrote:

I went up the other day behind the molino past that Moorish keep and there's a path through the live-oak forest to the top. It took me one hour and twenty minutes to get where I could look down on Valdemossa and across to Palma ... It was so sticky by the time I had finished with the olive-tree level and got into the oak level that I took off all my clothes except my shoes and stuffed them into my knapsack. But met no one.

R.R. Graves, in O'Prey, 1982, 210

In February 1994 we also met no one after we had climbed up beyond the Mallorcan teenager level, but the mountain air was cold and we kept our clothes on. Looking down on the geography of Deiá from above, it seemed for a moment to make sense. Or perhaps our transient sense of continuity with the past was merely nostalgia for the greater certainties of our own youth, when geographical reality had emerged in sharp focus from the pages of the books that we read.

At the end of that day there were more reminders of how limited are the insights that the bird's-eye-view can offer. In Deiá itself, at dusk, we walked up to the church, through silent streets, past empty, shuttered houses. Where were the owners? London, Frankfurt, California? In the tiny churchyard on the hilltop it was almost too dark to read what had been scratched in the wet cement of a plain concrete slab: 'Robert Graves, Poeta, 24.7.1895–7.12.1985'. There were other names on more elaborate memorials, English, Americans and Germans, as well as Mallorquin and Spanish names. Night was falling as we retraced our steps through the narrow streets, when suddenly our way was blocked by a silent mass of people. Twenty or thirty men were walking towards us with set faces and blank eyes. As they came closer we could see that none of them was young. They were wearing overcoats and jackets with the collars turned up against the cold wind. At the front there were four men in black suits balancing a coffin on their shoulders. We crouched in a doorway to let the funeral procession go past. No words were spoken.

We were in a hurry now to leave Deiá and the ghosts of Robert Graves. We decided not to visit El Olivo, the restaurant recommended in the guide ('Adventurous nouvelle cuisine, popular with the Mallorcan chic'). We drove instead to Puerto de Sóller, encouraged by the guide's description: 'The port is principally a tourist resort and restaurants are poor' (Hewson, 1990, 94). It was time for us to reconsider geographical reality in a more familiar setting. What we had seen that day signalled the growing gap between past and present, despite their inevitable co-existence. We had been tempted to see the present geography of Mallorca as a continuation of the past, an illusion made easier by our position as detached outsiders. We had been seeing the cultural landscape as if we were floating above it, interpreting it as a pattern on a map or as a model of social and ecological relationships. But every close encounter had shown that Mallorca contained dimensions which did not conform to map or model.

4.2 The New Geography: a power base for insiders

A few months after Linda McDowell and I returned from Mallorca, in the summer of 1994, there took place in Cambridge several reunions to mark the retirement of Professor R.J. Chorley. Inevitably, these occasions encouraged all of those present to reflect on the changes in geography over the course of his career. Dick Chorley's first decade in academia had been a period of extraordinary ferment in British geography, in which he and Peter Haggett (University of Bristol) took leading roles. During these years, either jointly or individually, Chorley and Haggett published *Frontiers* (1965),

Locational Analysis (1965), *Models* (1967), *Water, Earth and Man* (1969), *Network Analysis* (1969) and *Directions* (1973), all of them flagships of what came to be called the 'New Geography'. This term was, apparently, first used in a newspaper review of *Frontiers in Geographical Teaching* that came out in March 1966 (Stoddart, 1987, 335), just about the time when, as a newly arrived student, I began to think of myself as a geographer. In fact, the Cambridge department was by no means dominated by the New Geography in the late 1960s. Physical, historical and economic geography of a highly traditional kind co-existed uneasily with the new wave, which was mocked by Ben Farmer in his lectures as 'Haggettry'. But the traditional approach of people like Farmer seemed to offer us little beyond a disorderly and descriptive regional geography, whereas the alternative vision was a mega-structure of breathtaking dimensions. The New Geography was no less than a blueprint for the construction of a new nomothetic science of geography through systems analysis, model building and quantification.

An anecdote about David Stoddart may convey the flavour of the times. In his lectures, David was usually entertaining and inspiring, but in the winter of 1967 he also had the unenviable task of convincing the first-year practical class that statistical techniques were both useful and important. He obviously decided that a full-frontal assault was the best strategy, so he boomed out the following message. 'Geography is a science or it is nothing. Progress in science depends on the rigorous testing of hypotheses, which in turn requires data that can be collected and analysed in a systematic way. Only quantitative data meet these requirements. Therefore, in a very real sense, data that are not quantitative are *inferior data*'.

Years later, when I became a lecturer myself, David Stoddart gave me some advice: 'It's not really what you say, it's the way you say it'. In the statistics class in 1967 his message carried absolute conviction. I wrote down in my notebook 'qualitative data = inferior data', and whether or not we all wrote it down, most of us believed it. Otherwise I find it hard to explain the real enthusiasm with which we set about David's exercise, which involved countless measurements from maps of the drainage density of Viti Levu island, Fiji. In the weeks that followed we carried out test after test to see if there was any statistical difference between the forested catchments of the wet zone and the savanna grasslands of the dry zone. Never mind that it was tedious work involving some endlessly repetitive number-crunching, employing a sort of stone-age technology of mechanical calculators. We did it willingly because it was not just geography, it was science.

The rhetoric of the New Geography became more and more grandiose. In *Models in Geography*, published in my second year, Haggett and Chorley were telling us that the adoption of this quantitative, model-building, hypothesis-testing approach was nothing less than a new *paradigm* for our subject. They were using the term 'paradigm' in the same sense as Thomas Kuhn in his book *The Structure of Scientific Revolutions* (1962). A Kuhnian paradigm is a stable pattern of scientific activity. Those whose research is based on shared paradigms feel themselves committed to the same problems, rules and standards. The dominant paradigm both defines the importance of questions for study, and at the same time sets criteria for the acceptability of solutions. It enables highly focused work to be done within an established research tradition. However, the rules of the tradition are acquired indirectly through one's education rather than being formally taught. Only when the goals and procedures of 'normal science' no longer seem satisfactory, and only when a better alternative is seen to be available, does the moment of paradigm shift take place.

By accident I had stumbled into what seemed like the epicentre of seismic events in my chosen discipline. By the mid-1960s, Haggett and Chorley knew that the time had come to shake the whole edifice of geography by shifting its paradigm:

> *Geography, coming late to the paradigm race, has the compensating advantage that it can study at leisure the 'take-off' paradigms of other sciences. There is good reason to think that those subjects which have modelled their forms on mathematics and physics ... have climbed considerably more rapidly than those which have attempted to build internal or idiographic structures.*
>
> Haggett and Chorley, 1967, 38

To Chorley this meant the rejection of Davisian geomorphology and all those who were still promoting it through the study of denudation chronology. Instead we should be studying landforms as dynamic systems, through the detailed study of *process*. What did this mean in plain language? In the informal setting of his supervisions (in my year we were a tutorial group of four), Dick Chorley provided a graphic metaphor: 'We can only make progress by standing on the shoulders of those who have come before us, by building on what they have done. But, never forget that if you stand on someone's shoulders, you are very likely to shit on his head'. To shit or be shat upon? That seemed to be the choice. If there was further doubt in our minds, who could forget the example of W.M. Davis (this was another of Dick's favourite jokes), striving throughout his long career to explain landforms without understanding process, but now thoroughly involved in processes, 'six feet underground!'.

In short, what the New Geography was offering us was a small part in a kind of revolution. In the tranquillity of a small market town in May 1968, it was rather frustrating to be reading about student unrest all over Europe. In France it seemed that a student–worker alliance might bring down the Fifth Republic of General de Gaulle. There were barricades in the streets of Paris, but in the streets of Cambridge we knew that Part I of the Geographical Tripos was imminent, and the only revolution that really concerned us was one of geographical methodology. We were not going to change the world, but at least we might be able to change the way we studied it.

That summer, while hippies were wandering the streets of San Francisco with flowers in their hair, I spent some weeks crawling over the floor of a disused scout hut near my home in Brighton, sellotaping together Ordnance Survey maps and tracing on to them a vast pattern of hexagons with coloured felt-tip pens. This was my Part II dissertation, and I was mapping the spread in the West Midlands of foot-and-mouth disease among livestock. The conclusion that I wrote to that dissertation sums up perfectly the spirit of the times:

> The study also has some wider relevance, showing that an epidemic such as this can profitably be studied as a predominantly spatial phenomenon. Its behaviour has been shown to be subject to similar laws to that of other phenomena diffusing through time and space, and its analysis to be amenable to similar techniques and models. As such, the study is an indication that geography itself can have a potential relevance wider than that implied by its traditional role.
>
> Bayliss-Smith, 1969, 101

Some undergraduates still write like this, but today they have different ways to show that they have become insiders, 'proper chaps' (but some are now females), cognoscenti of the most up-to-date revelation of geographical reality. The confident style verges on blithe arrogance. There is a casual use of buzz-words (spatial, diffusing, laws, models, relevance), and the atmosphere is one of barely suppressed evangelical fervour. All these mark it as a distinctive product of the times.

One of my other supervisors, Brian Robson, later wrote:

> The people involved in the group in the 1960s were mostly self-taught statisticians who were caught up in the excitement of new skills, the thrill of the battle, and the tangible results that techniques gave. Those who followed in the 1970s began to develop distinctively spatial techniques and brought to their work a far higher level of sophistication . . .
>
> Robson, 1984, 112

In the 1970s I didn't do this. By then I had decided that spatial science was not all that interesting and, in any case, it was getting too difficult. What remained for me was an almost compulsive urge to measure, to count and to model, along with the conviction that the New Geography was a worthwhile enterprise, and one that my generation of research students was uniquely qualified to pursue.

In this respect the New Geography was no different from any other revolution. It became the means of advancement for a whole generation of those swept along in its train. None of us had then encountered Michel Foucault, or had considered his view that the genealogy of knowledge should be analysed in terms of tactics and strategies of power:

> *Science, the pursuit of truth, ... and ritualised procedures for its acquisition, have traversed absolutely the whole of western society ... What is the history of this 'will to truth'? What are its effects? How is all this interwoven with relations of power? If one takes this line of enquiry then such a method can be applied to Geography. Indeed, it should be ...*
>
> Foucault, 1976, 66

But even without Foucault to guide us, it was clear to everyone that the New Geography offered a means towards higher status for a discipline long regarded in most universities as an easy option. Years later, Stoddart (1987, 328) commented: 'The technical achievement has been remarkable; the advance in conceptual sophistication equally so. Had it not been so, needless to say, we would all have been out of business.'

For those who were trying to do it, it provided other benefits. There was a common sense of purpose and quite a number of shared techniques. I remember a seminar organized by the research students, probably in 1970, entitled 'Changing some Wheels on the Diffusion Band Wagon'. The seminar was attended by people in quite disparate fields of human, physical and historical geography. We were all on the New Geography bandwagon, and it was rolling along well.

Perhaps we should have paid more attention to the final words in Haggett and Chorley's opening chapter in *Models in Geography*. They warned that even if we are successful in building and testing models in geography, we should not expect any progress towards a *full* understanding of the interacting systems of humans and their environment – on the contrary, 'scientific effort does not reduce the sum total of problems to be solved, it rather increases them'. In other words, normal science solves a few puzzles, but also reveals deeper contradictions. Perhaps it is only at the crest of the wave that one can believe fully in the wonderful truths that are revealed by the new paradigm. T.H. Huxley's Stage II of public opinion, 'The Novelty is Absolute Truth', sums up this passing phase (Table 4.1). For the New Geography, this stage was ending by the time my generation of students was out in the world doing and teaching geography.

This is not the time or place to write a proper history of the New Geography. Was it a new paradigm? Yes, but the shift to it never involved more than a few geographers, and it was not sustained. Historical geographers like Jack Langton and Derek Gregory drifted off in other directions. Human geographers like David Harvey, Brian Robson and, to some extent, Peter Haggett himself, did likewise. David Stoddart came to reject the revolutionary rhetoric of the New Geography, and regarded 'paradigm' shift as an inaccurate description of how change actually takes place in the history of science. To him, the more interesting question is why the paradigm idea remains so popular. It seems possible that in geography those who propounded the Kuhnian interpretation of change were doing so in ways which tended to make it self-fulfilling (Stoddart, 1981, 78).

Table 4.1 *T.H. Huxley's four stages of public opinion, as supplied to the* New Geography

---

Stage I: Just after publication

    The Novelty is absurd and subversive of Religion and Morality. The propounder both fool and knave.

(New Geography, 1950–65?)

Stage II: 20 years later

    The Novelty is absolute Truth and will yield a full and satisfactory explanation of things in general. The propounder a man of sublime genius and perfect virtue.

(New Geography, 1965–75?)

Stage III: 40 years later

    The Novelty will not explain things in general after all and therefore is a wretched failure. The propounder a very ordinary person advertised by a clique.

(New Geography, 1975–?)

Stage IV: 100 years later

    The Novelty is a mixture of truth and error. Explains as much as could reasonably be expected. The propounder worthy of all honour in spite of his share of human frailties, as one who has added to the permanent possessions of science.

(New Geography, ??)

After Bibby (1959, 77) and Stoddart (1981, 77)

However, Stoddart remains a passionate advocate of what we might see, in retrospect, as the central message or theoretical 'high ground' of the New Geography. He shows that if we study geography in a holistic way, then we not only inherit just an important intellectual tradition, but we also have a powerful means for comprehending the growing problems of our degraded, culturally divided and over-populated planet (Stoddart, 1987). One example would be the conflicting 'explanations' that have been proposed for famine in less developed countries. Susan Owens and I have argued that to understand fully the causes of famine we have to apply several approaches simultaneously, and we must do this at a range of spatial scales. Existing models provide only partial explanations, and all make assumptions that in fact are ideological claims. The truth can only be approached by embracing methodological diversity (Bayliss-Smith and Owens, 1994).

4.3 The story so far

A summary of my underlying argument up to this point might run as follows:

1. Autobiography is not a straightforward task. To explain fully why one went about doing something in a certain way, whether it is 'doing geography' or 'fighting the Great War', involves a long and hard look at motives and values as well as influences. As Robert Graves found in *Goodbye to All That*, the task becomes easier if you have gone through a change in outlook, leading to a more detached and more self-critical assessment.

2. Although it helps us to see things more clearly, a detachment from the real world is also hazardous in ways that Robert Graves chose to ignore. To see the plains it is always tempting to climb the mountain. But the bird's-eye-view, in geography as in autobiography, encourages a selective approach to the evidence. It imposes a remoteness from reality that makes it easy to be seduced by surface patterns. Any closer look at a cultural landscape, as on our Mallorca field trip, reveals a startling mismatch between the surface appearance and the underlying processes. The surface shows many beguiling patterns inherited from the past, but there are hidden processes at work which contradict what is implied by these patterns, and which ultimately will transform them.

3. The ways in which this cultural landscape can be studied are always changing. Doing geography in a certain way does not emerge as the inevitable and logical consequence of a certain history of ideas. The choice of topic might emerge from private study, but it is more likely to be the suggestion of someone that one respects. The choice of method is to a large extent the consequence of social conditioning within a certain intellectual climate. Quantification and model building was, for British geographers in the early 1970s, the smart thing to do. It had become 'normal science' in those departments washed over by the diffusion wave of the New Geography. In the Cambridge department few resisted, and I was not one of them. Only through fieldwork did I start to question my intellectual conditioning, and so begin the process of redefining geographical reality.

4.4 Doing island ecosystems

Doing geography as an undergraduate had detached me from the real world so thoroughly that I found, after three years, that I had formed no ambitions beyond wishing to continue doing it. After completing my B.A. in 1969, I went to South America for three months with a student expedition. On my return I drifted back to Cambridge. Letters had gone astray and no one really expected me back, but fortunately a Social Science Research Council studentship was available and David Stoddart was prepared to act as a supervisor. The project that I had in mind seems rather strange to me now, but perhaps no more strange than some of the things that I had been doing already, like measuring the drainage density of Viti Levu or plotting outbreaks of foot-and-mouth disease in Cheshire. I was still keen on measurement, but this time it was people.

I did some reading on the subject of islands and ecosystems, and within a few weeks I had drafted a three-page 'Outline of Proposed Research'. I followed this up with a discussion of what 'variables' needed to be measured, and what kind of island might be suitable. At that time nobody's research plan was subjected to the kind of public scrutiny that the proposals of graduate students receive these days. As Brian Robson remarked about his own first steps in research, 'it seems unlikely that a postgraduate today would be allowed the same degree of luxury to meander quite so freely and with so few

> *The economies of human communities can usefully be thought of as systems ... The functioning of these ecosystems can best be analysed by measuring the energy flows between their different elements. In practice this procedure has only been carried out successfully for simple biological systems. This approach has also been used by some anthropologists and geographers, but their treatment of man in the ecosystem is either partial or merely descriptive ... In order to measure fully all these variables and hence produce satisfactory results, a human system sufficiently discrete and simple must be found. Small islands suggest themselves at once as especially suitable ...*
>
> (unpublished, 1969)

constraints on what was tackled' (Robson, 1984, 108). For me the important thing was that my supervisor was happy with my proposal and was prepared to support my various applications for travel grants.

What exactly was I proposing? Some of the language of my proposal now seems a little peculiar. The reference to 'man in the ecosystem' was not intended to exclude women, and was probably an unconscious reference to the book *Man's Place in the Island Ecosystem* which I had been reading (Fosberg, 1963). What is more interesting, and more revealing, are the assumptions that I made about how one should study geographical reality. I laid them out in a quite explicit way but never thought to question their ideological basis, which was nothing less than a total commitment to the methodology of positivist science.

The first assumption was that satisfactory results depend upon being able to measure fully all the variables ('qualitative data = inferior data'). The second was that measuring people/environment relationships in terms of energy flow is the 'best' approach. I had become convinced of this through reading David Stoddart's chapter in *Models in Geography* entitled 'Organism and ecosystem as geographical models'. I have just re-read this chapter and I still find its arguments quite persuasive, but I now notice in the text some warning notes which, as a graduate student, I was happy to ignore. Stoddart warns that the emphasis by biologists on the energetics of ecosystems 'is clearly of peripheral geographic significance'. What is more fundamental, he says, is the concept of *systems* in geography, which was going to be 'central to the development of the subject as a nomothetic science' (Stoddart, 1967, 537).

The most impressive ecosystem research that had been published at that time was the study by H.P. and E.T. Odum of a coral reef in the Marshall Islands, and J.M. Teal's study of a salt marsh in Georgia (Figures 4.1, 4.2). These reports were the outcome of years of work by large teams of researchers. My intended team consisted of me, a rather reluctant wife and, assuming that I could find one, a part-time field assistant. What kind of system should I choose? My supervisor had some suggestions:

Most ecosystems involving man are more complex than the salt-marsh and coral-reef systems already described, and attempts to describe ecosystems at such complex levels are likely to be difficult until experience is gained with relatively simple or restricted systems. Fosberg's focus on islands is one way out of the problem; another ... is to concentrate on primitive human and sub-human groups ...

Stoddart, 1967, 529

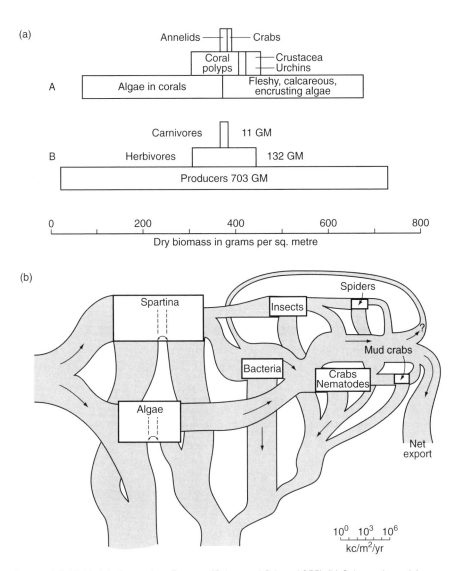

Figure 4.1 *(a) Model of a coral reef system (Odum and Odum, 1955) (b) Salt marsh model (Teal, 1962).*

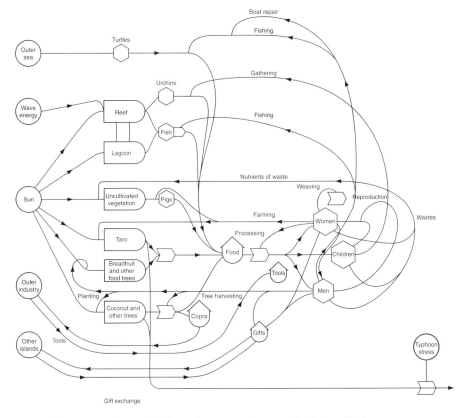

Within the figure:

Boat repair
Fishing
Outer sea
Turtles
Gathering
Urchins
Wave energy
Reef
Fishing
Fish
Lagoon
Nutrients of waste
Weaving
Reproduction
Sun
Uncultivated vegetation
Pigs
Wastes
Farming
Women
Taro
Processing
Food
Children
Breadfruit and other food trees
Tools
Outer industry
Planting
Tree harvesting
Men
Coconut and other trees
Copra
Other islands
Tools
Gifts
Typhoon stress
Gift exchange

Figure 4.2 *More complex model of a reef system involving people (Odum, 1971).*

   Stoddart went on to discuss 'the intriguing possibilities of primate geography', but I could never quite be sure when he was being serious. I drew the line at gorillas, but a primitive island sounded quite promising. Something Polynesian perhaps? I was much encouraged by reading a book called *Pigs for the Ancestors* by an American anthropologist, Roy Rappaport (1968), who had lived for two years in a village in the New Guinea Highlands. The book included an analysis of energy flows in the community, which showed that as time passed more and more food and work energy was directed towards the growing population of pigs, culminating in a crisis which triggered the ritual slaughter of the herd and the beginning of a new religious cycle. As with sacred cows in India, there seemed to be good ecological reasons for ritual practices: the ecosystem's relentless logic had reduced religion to a mere 'noise' factor. As an atheist about to marry a Roman Catholic, I found this rather gratifying.

   This New Guinea study was also encouraging in another way. One reason that Rappaport was so interested in food energy was because his first degree had been in Catering and Hotel Management, before he switched to Anthropology. A degree in Geography was perhaps not such a bad qualification. It was obviously time for me to start making enquiries.

In early December I received the following reply to a letter I had written to the Office of the District Commissioner in Auki, British Solomon Islands Protectorate:

> *Dear Sir, Please refer to your letter of 17ᵗʰ November, 1969, about your proposed fieldwork at Sikaiana or Ontong Java. I have referred your request to the Chief Secretary and asked him to reply to you direct. It is unlikely there would be any objections from the people of either island, but I agree that Ontong Java would probably be the more rewarding. There are two villages, Leuaniua and Pelau, about 40 miles apart at the southern and northern ends of the atoll with a combined population of about 800–900 people . . . I am afraid that at this stage I cannot accurately forecast the dates of shipping . . .*
> *Yours faithfully,*
> *Richard Turpin, District Commissioner*

There was a further letter on 20th January:

> *During the forthcoming election ships will be leaving Auki . . . I have no doubt that passages could be found for you and your wife on one of these voyages. A District Officer will be visiting Ontong Java in February – he could, if you would like him to, arrange for accommodation for you – do you want a leaf house built and if so what sort of size and what funds do you have?*

By 10th March our accommodation had been arranged:

> *[Your house] has a wooden floor, leaf walls and roof and consists two adjoining rooms and a small verandah. Any necessary repairs will be made and you will be charged a rent of $4 a month, one half being paid to the landowner and the other half to the owner of the house itself. I think this would in fact be cheaper for you than building a new house . . .*

Two months later the government boat landed us at Luangiua, Ontong Java atoll, together with our luggage, food supplies, kerosene lamps, notebooks, camera film, plant presses, polythene bags, collecting jars and preserving fluids, formalin and ethanol. As we had been promised, the house was ready, and the following day I was able to make my first scientific observation:

- June 6th 1970
- Maximum temperature: 32.5° C
- Minimum temperature: 28° C
- Wind: SE, Force 5
- Rainfall: 2 mm

At last I was doing geography.

4.5 Robinson Crusoe's folly

In the Age of Discovery, the geographer was often represented as a man contemplating a globe, but for most of us today a more apt image would be of an investigator contemplating an exotic landscape (glacier, rainforest, Third World village, city street) and collecting data. The actual process of data accumulation is becoming more and more streamlined, but its underlying purpose remains the same. Feeding data into a laptop computer is no different from making field notes or even carving a notch on a post to mark the passage of time. In each case, the act of recording symbolizes the position of the scientist as a detached observer of reality. My daily weather notes can be deconstructed in this way. I had no particular interest in the weather on Ontong Java, but the mere act of recording it signified that I was not a tourist, nor a beachcomber, nor an indigenous islander. I was like Robinson Crusoe, a Western intruder in the island world of Man Friday, but not someone with any intention of abandoning his culture or 'going native'.

In the character of Robinson Crusoe, Daniel Defoe has provided us with an almost archetypal symbol of cultural intrusion and ethnocentrism:

> *After I had been there about Ten or Twelve Days, it came into my Thoughts that I should lose my reckoning of Time for want of Books and Pen and Ink, and should even forget the Sabbath Days from the Working Days; but to prevent this I cut every Day a Notch with my Knife upon a large Post ...*
>
> Defoe, 1719, 66

Crusoe also found that keeping a journal had therapeutic value: it helped 'to deliver my Thoughts from daily poring upon them, and afflicting my Mind'. For me, writing letters back home fulfilled the same objective. I still have my rough drafts of those letters.

> *24 June 1970*
>
> *Dear Dr Stoddart,*
>
> *We have been here nearly 3 weeks now, and are settling in quite well. We have a small but comfortable leaf house on the edge of the village near the ocean side, so I am taking this opportunity of letting you know what is happening. The general picture is potentially quite promising. The community here at Luangiua seems flourishing and quite balanced. The subsistence economy is virtually intact ... All this I hope to quantify in terms of activity patterns ... I have also run a preliminary food survey (intake of 6 households for 7 days) ... Taro provided on average 77% of the calories ... As a side interest I have been doing some birdwatching and have already added several species to the record list ... I should be very grateful for your comments, and any news of the Department.*
>
> *Yours, Tim*

My supervisor cannot have been interested in the arcane details that I poured out in this five-page letter, the first of many that I wrote. However, as an experienced castaway himself, he must have realized the therapeutic value of Robinson Crusoe's journal and, to his credit, he always replied.

Isolated islands merely accentuate the self-doubt and discouragement that are major hazards of any fieldwork. When you are confronted with something quite different from the geographical reality you had so carefully constructed back in the department, it can be hard to sustain belief in the paramount importance of doing geography. Outside the fevered atmosphere of the university, the project that once seemed viable and exciting begins to appear foolish or even pointless. In what terms could I explain my fieldwork to the inhabitants of my chosen island? I could speak with many of them in Melanesian Pidgin, but to explain my project was difficult not for linguistic reasons, but because its agenda only made sense in terms of an esoteric paradigm. Instead of explaining my project, it was easier to adopt the role of harmless eccentric. The Ontong Java people must have found this rather strange. The white men they had previously encountered were either traders, missionaries or district officers, and I seemed to fit into no known category.

I was very fortunate to find two knowledgeable field assistants, David Kaiaenga and Timo Kepangi, who also became good friends. At times they appeared to have more faith in my agenda than I did myself. My supervisor had also given me some excellent advice: record everything, collect everything, publish everything. Every letter that David wrote urged me on:

> I had a letter from Ray [Fosberg] about you about ten days ago; he is most interested in the vegetation and flora of Ontong Java, and reckons it is close enough to the Solomons to pick things up. You should hear from him soon . . .
>
> Incidentally, we have been looking over the Solomons corals in the BMNH [British Museum of Natural History] . . . The only corals from the Solomons in the BM are those collected by Guppy in the 1880s . . . Two or three afternoons collecting would be well worthwhile if you can find the time . . .
>
> I have talked to R.M.S. Perrin of the School of Agriculture about the taro soils. He is prepared to organise something in the way of analysis of specimens when you return . . . The other thing in your letter was about sets of plants. I think you need to collect multiple sets without any doubt . . . Fosberg will retain a set for the U.S. National Herbarium . . .
>
> Any cargo cults on Ontong Java? I bought Worsley's The Trumpet shall Sound to read on the train the other day – it is now in paperback – and became quite engrossed.
>
> All good wishes to yourself and Francoise for Christmas. Don't have too much roast pig.
> Yours ever, David

By and large I followed his advice, and the days were therefore filled with purposeful activity. I collected specimens of plants and fish and corals; I mapped the villages, the taro gardens, the coastal changes; I counted birds and houses and canoes; I recorded legends, genealogies, and women's reproductive histories. In the years that followed I was able to write articles about Ontong Java birds, population change, history of European contact, canoes and voyaging, tattooing, village economy, hurricanes and landforms, fish and fishing. There are papers still to be written on the vegetation, the coral reefs and the settlement pattern. But all of these observations were incidental to my main purpose, that of constructing and quantifying an energy flow model of the relationship of the people to their island ecosystem.

As the weeks went by, this central task, my PhD topic, began to seem more and more dubious, but it kept me very busy. It involved an immense programme of population censuses, activity surveys, time-and-motion studies, land use mapping, yield studies, and the monitoring of imports and exports. I needed samples of households, samples of area, samples of time. Everything that moved I counted, or measured, or weighed. When I made a return visit to Ontong Java 15 years later, in 1986, the women would point at me, laughing, and explain to their children that this was the man who had stood and waited in the hot sun, day after day for weeks, simply in order to weigh their baskets of taro. Not just one woman's basket, but everyone's. In retrospect I could join in their laughter, but at the time it wasn't so funny. My project began to seem ludicrous, even to me. What was this model that I was so obsessed with? And would it ever get me off the island?

My model began to seem rather like the boat that Robinson Crusoe made out of a huge tree trunk. For weeks and months he hacked and hewed with axe and hatchet and 'inexpressible Labour':

> *I pleas'd my self with the Design, without determining whether I was ever able to undertake it; not but that the difficulty of launching my Boat came often into my Head; but I put a stop to my own Enquiries into it, by this foolish Answer which I gave my self,* Let's first make it, I'll warrant I'll find some way or other to get it along, when 'tis done.
>
> Defoe, 1719, 126

In the end, all of Crusoe's efforts to drag his boat to the sea were in vain, and he saw at last 'the Folly of beginning a Work before we count the Cost'. What he had intended as the means for his escape from the island had led him instead to the edge of despair.

I also laboured for months on constructing my model, but it remained as static as Crusoe's boat. By the end of the year I had calibrated its inputs and outputs, and I could track its internal workings with some precision. I knew how much taro and fish were produced, how much sugar, rice and kerosene were purchased, how much copra was sold. I knew the work inputs needed to generate all these outputs. I could trace the pattern of energy flow. I had measured almost everything, but increasingly I came to believe that I had understood almost nothing. The small aspects of island life where I did feel that I had achieved some understanding were the result of casual observations, chance conversations and anecdotes. There hadn't been much time for this ethnographic type of fieldwork, because of my heavy programme of counting and measuring and mapping. Besides, qualitative data were inferior data. Surely with all my numbers I could successfully launch my model?

The more I looked at the island ecosystem model, the less it was able to tell me what I didn't already know. It described in a generalized way what was happening on the atoll, but not what was going to happen next. By reducing people to mere transformers of energy, I had removed from the model any information about people's needs, desires and aspirations, beyond the minimization of effort and the satisfaction of present needs. If everything I had assumed about the present did not change in the future, then I could predict when a crisis point would be reached because the atoll's carrying capacity had been exceeded. But, in fact, everything was changing, and the atoll was not a closed system. New aspirations and opportunities were beyond the scope of my study, so my prediction of an impending energy crisis was just another doomsday scenario. My twelve months as a castaway were coming to an end, and there was no one to rescue me except myself.

To cut a long story short, 'normal science' came to my rescue. I had embraced the New Geography paradigm, and in the field, within my chosen corner of geographical space, I had followed faithfully all the procedures that the new paradigm involved. It had provided the rules which told me what the world was like. These same rules had allowed me to concentrate on an esoteric problem that the rules, together with existing knowledge, had defined. And, as Haggett and Chorley had predicted, the paradigm had turned out to restrict me rather than liberate me:

> Paradigms tend to be, by nature, highly restrictive. They focus attention on a small range of problems, often enough somewhat esoteric problems, to allow the concentration of investigation on some part of the man–environment system in a detail and depth that might otherwise prove unlikely, if not inconceivable.
>
> Haggett and Chorley, 1967, 27

My fieldwork was complete, and back home little had changed. Systems still ruled. My energy flow model was not greeted with ridicule, and even world events had moved in my favour. When I came to complete my PhD research in the winter of 1973–74, there were strikes among the miners and electricity workers, leading to power cuts. Parts of the last chapter of my dissertation were written by candlelight. The oil price had been quadrupled by OPEC countries and the newspapers were full of stories of an impending 'energy crisis'. My obsession with energy flows in the ecosystem was beginning to seem less esoteric, and my data took on some new relevance. It passed the scrutiny of my PhD examiners, and later some was even published (Bayliss-Smith, 1977, 1982). However, by then I had moved towards a different paradigm for doing geography. I was in Viti Levu, Fiji, but instead of doing ecosystems I was doing development – but that is another story.[3]

Postscript

Obviously I regard my personal quest to define the geographical reality of island ecosystems as an unfinished journey. In retrospect, I believe my starting point for this journey was not helped by the ambitions of the New Geography to integrate the physical and human parts of the subject in a particular way. I have presented my own doomed attempts to apply this paradigm by modelling island/islander relationships through energy flows as a sort of cautionary tale. I understand that physical geographers still believe that modelling environmental systems is a worthwhile endeavour, but perhaps today no one would extend this approach to embrace the various fallacies that I have explored in my own story.

But the dangers remain. I attended in December 2001 a conference in Samoa on 'Adaptations of Pacific Islanders to Climate Change and Climate Variability'. Once again I became aware of the imaginative appeal of a new paradigm (global warming), and how easy it is for outsiders (even Pacific islanders) to be swept along by the persuasive rhetoric of the 'proper chaps', in this case climate modellers. Physical geography now has powerful tools at its disposal, and many wonderful opportunities to contribute to our understanding of the planet. The lessons we can learn from the New Geography show, I believe, the need for those engaged in environmental modelling to recognize its limitations as well as its strengths.

References

Bayliss-Smith, T.P. 1969: 'An analysis of the spatial diffusion of the epidemic of foot and mouth disease in Britain, 1967–68'. Dissertation for Part II of the Geographical Tripos, Cambridge University (unpublished).

Bayliss-Smith, T.P. 1977: 'Energy use and economic development in Pacific communities'. In T.P. Bayliss-Smith and R.G.A. Feachem (eds.), *Subsistence and Survival: Rural Ecology in the Pacific*. Academic Press, London, 317–359.

Bayliss-Smith, T.P. 1982: *The Ecology of Agricultural Systems*. Cambridge University Press, Cambridge.

Bayliss-Smith, T.P. and S.E. Owens 1994: 'The environmental challenge'. In D.J. Gregory, R.L. Martin and G.E. Smith (eds.), *Human Geography: Society, Space and Social Science*. Macmillan, London, 113–145.

Bibby, C. 1959: *T.H. Huxley: Scientist, Humanist and Educator*. Watts, London.

---

3 I must thank all those teachers, friends and colleagues mentioned here, who have helped me on my journey towards some inkling of geographical reality. Others are also moving silently through these pages. As Robert Graves wrote about Laura Riding, 'Yet the silence is false if it makes the book seem to have been written forward from where I was instead of backward from where you are'.

Defoe, D. 1719: *The Life and Strange Surprizing Adventures of Robinson Crusoe.* William Taylor, London. Reprinted 1983, Oxford University Press, Oxford.

Fosberg, F.R. (ed.) 1963: *Man's Place in the Island Ecosystem.* Bishop Museum Press, Hawaii.

Foucault, M. 1976: 'Questions à Michel Foucault sur la géographie'. *Hérodote* 1, Paris. Reprinted in C. Gordon (ed.) 1980: *Michel Foucault, Power/Knowledge: Selected Interviews and Other Writings 1972–1977,* Harvester Press, Hemel Hempstead, 63–77.

Graves, R. von R. 1929: *Goodbye to All That.* Jonathan Cape, London.

Graves, R. von R. 1930: *But it Still Goes On.* Jonathan Cape, London.

Graves, R.P. 1990: *Robert Graves: The Years with Laura.* Weidenfeld & Nicolson, London.

Haggett, P. and Chorley R.J. 1967: 'Models, paradigms and the New Geography'. In R.J. Chorley and P. Haggett (eds.) *Models in Geography* Methuen, London, 1941.

Hewson, D. 1990: *Mallorca: The Epicure's Guide.* Merehurst Press, London.

Kuhn, T.S. 1962: *The Structure of Scientific Revolutions.* University of Chicago Press, Chicago.

Odum, H.T. 1971: *Environment, Power, and Society.* John Wiley, Chichester.

Odum, H.T. and Odum E.P. 1955: 'Trophic structure and productivity of a windward coral reef community on Eniwetok Atoll' *Ecological Monographs,* 25, 291–320.

O'Prey, P. (ed.) 1982: *In Broken Images. Selected Letters of Robert Graves, 1914–1946.* Hutchinson, London.

Rappaport, R.A. 1968: *Pigs for the Ancestors: Ritual in the Ecology of a New Guinea People.* Yale University Press, New Haven, Connecticut.

Robson, B.T. 1984: 'A pleasant pain'. In M. Billinge, D. Gregory and R. Martin (eds.) *Recollections of a Revolution: Geography as Spatial Science,* Macmillan, London, 104–116.

Stoddart, D.R. 1967: 'Organism and ecosystem as geographical models'. In R.J. Chorley and P. Haggett (eds.), *Models in Geography.* Methuen, London, 511–548.

Stoddart, D.R. 1981: 'The paradigm concept and the history of geography'. In D.R. Stoddart (ed.), *Geography, Ideology and Social Concern.* Blackwell, Oxford, 70–80.

Stoddart, D.R. 1987: 'To claim the high ground: geography for the end of the century'. *Transactions of the Institute of British Geographers* NS 12, 327–336.

Teal, J.M. Jr. 1962: 'Energy flow in a salt marsh ecosystem of Georgia'. *Ecology,* 53, 614–624.

PART III

RESEARCH MEANINGS

5
The natural science of geomorphology?

Chris Keylock

> Pooh went into a corner and tried saying 'Aha!' in that sort of voice. Sometimes it seemed to him that it did mean what Rabbit said, and sometimes it seemed to him that it didn't. "I suppose it's just practice," he thought. "I wonder if Kanga will have to practise so as to understand it."
>
> A.A. Milne, *Winnie-The-Pooh*, 1926, 84

> A main source of our failure to understand is that we do not command a clear view of the use of words ... Our grammar is lacking in this sort of perspicuity.
>
> Wittgenstein, *Philosophical Investigations*, 1958, 122

5.1 Introduction

The scientific status of geomorphology, relative to the sciences of chemistry and physics, has been an ongoing point of discussion within the literature. An implication of the study by Bauer (1996) is that to those outside the subject, the scientific status of geomorphology depends upon the sub-discipline's host department:

> In the eyes of the public, geology is a science and geography is about maps, capital cities and longest rivers. In the eyes of the physicist or chemist, geology is a 'soft' science and geography is about maps, capital cities and longest rivers. In the eyes of the economist or sociologist, geology is a 'hard' science and geography is sometimes useful.
>
> Bauer, 1996, 399

Thus, there are societal and institutional issues that affect the perception of geomorphology's status within academia. However, if Bauer's remark is felt to be a reasonable summary of popular opinion, then from the perspective of physics and chemistry, geomorphology is at best a soft science, somehow different in nature to the 'hard' sciences. To investigate why this difference exists requires one to delve into methodological and philosophical issues. Two prominent (and related) reasons for this difference in status are commonly put forward in the geomorphic literature.

The first view was elegantly stated by Simpson (1963) and is still considered to be of critical importance (Frodeman, 1995). This is that history is an essential dynamic in earth science, whereas it plays a much less important role in classical physics and chemistry. From this it follows that the earth sciences must develop alternate methodologies, and therefore, potentially different ontologies in order to effectively study geomorphic systems. The second view is related, but focuses on the process of closure that is necessary in order to apply techniques from the 'hard' sciences to geomorphology. Invoking *ceteris paribus* clauses to simplify a landscape system, to the extent that it is amenable to mathematical and experimental analysis, implies that this component of the system is the core, which in reality is obscured by the noisiness of real-world processes and measurement. However, an alternative perspective is to consider the landscape system as embedded within a complex environment (with a complex history) and to prioritise the study of this system in itself. In other words, the 'noise' does not exist, it is an integral part of the system studied. This implies a very different commitment to the way the world works, which in turn leads to a difference in philosophical perspective between the 'hard' and 'soft' sciences. It was an argument along these lines that led Richards (1990) to advocate that geomorphology move away from approaches founded within a critical-rationalist or logical-positivist perspective and embrace a new outlook, termed 'transcendental realism' by Rhoads (1994).

My aim in this paper is not to engage with these debates, but to explore another possible reason for the difference in status between geomorphology and physics/chemistry. My argument centres on the concept of natural kinds, introduced into the geomorphic literature by Rhoads and Thorn (1996) and Rhoads (1999). The latter suggests that a more detailed investigation into geomorphic natural kinds might, 'prove fruitful for exploring the philosophical basis of classification in physical geography' (Rhoads, 1999, 766). Furthermore, Rhoads claims that the naming of landscape objects as particular landforms has resulted in an elaborate taxonomic system within geomorphology. This is an important issue that demands further attention. The argument I wish to make is that the geomorphic objects that are known to the everyday language community take their names from words in everyday usage. Even if the reference for the two terms differs, this affects the perception of the discipline. Indeed, within geomorphology this can lead to definitions that confuse scientific intent with everyday usage. This highlights the complex relations between geomorphology and a wider society as well as between geomorphology and the other sciences. Some consequences of the views put forward here are an anti-realism towards the definition of geomorphic objects and a relativistic perspective on the evolution of scientific ideas that is rather different to the general views held by Kuhn (1962), Lakatos (1970) and Feyerabend (1970).

However, it is at first necessary to set out what is meant by a natural kind. Hacking (1990) has called into question the utility of the term altogether, preferring 'scientific kind' and 'scientific term'. Although this view has some merit, in this article I stick with the conventional terminology. Initially, I focus my attention upon the well-known formulations of natural kinds developed by Kripke (1972) and Putnam (1975a, b, c), which imply a causal theory of reference and re-awaken the Lockean notion of 'essence' (Locke, 1690).

5.2 Natural kinds

Kripke (1980) commences his collection of lectures, *Naming and Necessity*, with an account of the properties of proper names. From this discussion he states that, '*names are rigid designators*' (Kripke, 1980, 48) and goes on to explain this by saying:

> ... *a designator rigidly designates a certain object if it designates that object wherever the object exists; if, in addition, the object is a necessary existent, the designator can be called strongly rigid. For example, 'the President of the U.S. in 1970' designates a certain man, Nixon; but someone else (e.g. Humphrey) might have been the President in 1970, and Nixon might not have; so this designator is not rigid.*
>
> Kripke, 1980, 49

It is his transfer of the idea that proper names are rigid designators to the view that all names are rigid designators that is important here, because this structures his account of natural kinds (objects that exist, are of explanatory importance and can be placed in a particular group based upon their intrinsic properties, independent of the form of classification we adopt). Kripke's argument for the existence of the natural kind of 'tiger' is as follows: 'We can say in advance that we use the term "tiger" to designate a species, and that anything not of this species, even though it looks like a tiger, is not in fact a tiger.' (Kripke, 1980, 121) That is, he explicitly rejects a 'cluster concept' or dictionary approach to fixing the reference for the term 'tiger' where a list of attributes (carnivorous, four legs, yellow-orange with black stripes, etc.) is sufficient to define the characteristics of the natural kind. As he points out, one is not necessarily confused by the statement, 'look, a three-legged tiger', and therefore, having four legs cannot be an essential part of the definition of 'tiger'. Given that the same is true for the other descriptive attributes of a tiger mentioned here (e.g. white tigers exist), the reference of tiger cannot be fixed in this manner.

Another important aspect of Kripke's argument, and something that Putnam (1975b) picks up on, is his concept of a 'chain of communication' that links together the people who use a term in a particular way, or who use a particular reference for a natural kind. Thus, these various communities play a part in what Kroon (1985) calls *reference-transmission*, and this is something that I will return to later in this article.

Hilary Putnam's opinion on reference fixing is rather similar. He opens his paper 'Explanation and reference' (Putnam, 1975b) with an example from Engels of how it would be weak science to consider a lungfish not to be a fish purely because it lacks gills. Putnam makes two important comments upon this example that show some of the difficulties in trying to use simple analytic definitions to fix the behaviour of a natural kind word.

(a) A 'cluster concept' definition of a natural kind might not be isomorphic to the actual natural kind, but it might *correspond* to it.

(b) The concepts that supply the reference for a natural kind are likely to change through time due to scientific discoveries etc., but the essence of the kind itself may not alter.

To try and deal with some of these complexities, Putnam (1975a) introduces the concept of 'stereotype'. For a specific natural kind, this is the *typical* case and is defined by a simplified theory of the properties of this natural kind. In addition, the *extension* (the set of things that the stereotype is true of) must also be known, or must be able to be determined by a sub-set of users of the given language, so that the stereotype (a linguistic construct) can be applied in the real world. These 'experts' can then be consulted to discover if an object *x* belongs to natural kind *y* or not. According to Putnam, it is important that only a sub-group is able to determine the extension of the term, otherwise the extension might become part of the stereotype, which would cause a change in meaning of the natural kind term. However, other philosophers would suggest that this change of meaning can occur and this view is similar to that defended in this paper. For example, Wittgenstein (1958, 79) states, 'The fluctuation of scientific definitions: what to-day counts as an observed concomitant of a phenomenon will tomorrow be used to define it.' Furthermore, Feyerabend (1958, 166) asserts that '*as a matter of fact* scientists re-interpret their observation-language L as soon as a new theory is devised which has consequences within L.'

Putnam (1975c) introduces a now-famous example to illustrate his ideas on natural kinds – the Twin Earth thought experiment. Twin Earth (W_2) is very nearly identical to earth (W_1), but with some important differences, one of which is that there is no H_2O, but instead a substance that is indistinguishable from H_2O at room temperature and pressure, but with a very different chemical formulae XYZ.[1] From Kripke's work it follows that even though both substances might be called 'water' in their local vernacular and have the same functioning in society and nature, they do not belong to the same natural kind. Putnam accepts this view explicitly, while also introducing an alternate theory for the meaning of water:

---

1 Episode 2F03 of *The Simpsons* makes a related point, but this time for the kind 'rain' as opposed to 'water' itself. In this episode, Homer travels to a twin world where his children are well-behaved and he owns a luxury car. He thinks this is just about perfect and asks Marge for a doughnut. She turns around and replies 'Doughnut? What's a doughnut?' On hearing this, Homer screams and immediately time-travels again. After he has gone, Marge looks out the window to sees doughnuts falling from the sky. 'Hmph. It's raining again,' she says.

> 1. One might hold that 'water' was world-relative but constant in meaning (i.e. the word has a constant relative meaning). In this theory, 'water' means the same in W_1 and W_2; it's just that water is H_2O in W_1 and water is XYZ in W_2.
> 2. One might hold that water is H_2O in all worlds (the stuff called 'water' in W_2 isn't water), but 'water' doesn't have the same meaning in W_1 and W_2.
>
> Putnam, 1975c, 231

Thus, both Putnam and Kripke hold that the latter is the correct theory and adopt a realist view of essence. The essential property of gold is the atomic number 79 and not, for example, its colour or ductility. However, I wish to argue that the former provides greater insight into our understanding of how language is linked to the objects we study, how this changes through time, and the relation between scientific and normal language use. That is, an anti-realist account regarding essences is advocated.

The account of Putnam and Kripke seems to pre-suppose an external position to evaluate the verisimilitude of our classifications of natural kinds, if their definitions are to prove useful in everyday and scientific languages. According to Putnam and Kripke, a natural kind is the same in all possible worlds, but surely our evaluation of whether or not a particular object is one natural kind or not is dependent upon the ability of our science to come to the correct conclusion regarding the determination of essence? Because our science is not perfect, we might be wrong, but we only know that we are wrong after the fact. We can never get outside of contemporary science to determine if we have the true essence of a natural kind. Thus, because we do not know what future science will consider to be the essential properties, we have to *believe* we are correct in our classification.

Imagine a situation where a geomorphologist labels an object as 'mountain', only to discover at a later date (with a change in scientific perspective or new instrumentation perhaps) that it is not a 'mountain', but a 'volcano'. According to Kripke and Putnam's approach, this geomorphologist was originally mistaken – it always was a volcano but they had failed to classify it appropriately. They had mis-identified the real essence of what it means to be a mountain or had defined the wrong set of objects (the wrong extension) for the natural kind 'mountain'. However, what would happen if, further into the future, this landform was believed not to belong to the natural kind 'volcano', but another natural kind x_k? This would mean that we today are wrong to use the term 'volcano' for this landform and should in fact call it x_k. But how are we to know this? What does it mean to be wrong in circumstances such as this? How does it affect our behaviour? Unless we know there is a 1:1 relation between our classification and reality, how do we attempt any classification?

I wish to argue that a particular word that designates a natural kind can mean different things to different social groups (or communities of speakers). Hence, our perspective determines what we believe to be the essence of a natural kind and a change in meaning of a natural kind term is difficult to distinguish from a change in the natural kind itself. This view is similar to the ideas of Kuhn (1990) and Hacking (1993) regarding changes in taxonomies and paradigm changes. Thus, the ontological commitment that Kripke and Putnam make to an external reality obeying time-transgressive laws is not questioned here. Rather, it is the assumption that natural kinds with fixed essences are an accessible part of this reality, instead of being part of the language-game of a particular community. Dupré (1981) and Wilkerson (1993) have highlighted some of the differences between ordinary language classifications of objects and taxonomies that would exist if everyday language was based upon natural kinds. The implication of this is that not only is there a difference between stereotype and essence, but this difference can lead to the formulation of rather different classificatory schemes. If this occurs, the meaning of a natural kind term changes, not only due to the difference between stereotype and essence, but in relation to other terms in the lexicons of particular social groups.

I believe that Putnam's division of linguistic labour between stereotype and extension is a very useful one, and a modified version of it forms an important part of my argument in this paper. As mentioned above, the chain of communication that links people's meanings is important for reference transmission. Putnam's division of linguistic labour can be used to highlight this.

Putnam states that a natural kind has a fixed essence. From the perspective outlined in this paper, where reference is *de dicto* and not *de re*, an *essence* for a natural kind must exist at a given time, but it can alter as the meaning for a particular natural kind itself changes. I wish to add the term *essence* to the discussion of the linguistic division of labour explicitly. Thus, if the extension is the set of objects which belong to the natural kind, or may be determined to do so, the *essence* is that property (or set of properties) that specialists use to determine if an object belongs to a natural kind or not (i.e. to determine the extension of the kind). Putnam defines the stereotype as the facts an everyday speaker needs to know in order to be able to use a term for a natural kind appropriately. Here, the *essence* is considered to be the state-of-the-art reference fixer for the natural kind that often will be known by far fewer people than the stereotype. However, even if a member of the public is aware of both the stereotype and the *essence*, there is still a role for the scientist, who has the tools and detailed knowledge to determine the extension for this case (as well as perhaps helping to refine the definition of the *essence* in the light of new theories).

If I am engaging in conversation with another non-specialist about an object y, which we believe to be a member of natural kind y_k, and for which we are only familiar with the stereotype (y_s), and we understand each other,[2] then as far as we are concerned, the

---

[2] Evidence for mutual understanding could arise if one of us finished the other's sentence or our behaviour was witnessed by a third person to be in accordance with our conversation – it showed that we understood the same stereotype and that we were engaged in the same language game.

object y is a member of y_k. Of course, a specialist (an extension definer) might come along and say that we are wrong and really y is a member of x_k. Perhaps, recent scientific research has shown that the *essence* of y_k is no longer y_e but x_e, causing y to be re-classified as a member of x_k. However, amongst her colleagues, this specialist may engage in lively debates about specific technical issues relating to the nature of x_e (perhaps the *essence* of x_e itself) or the manner in which the extension for x_k is determined, given a particular viewpoint of what x_e is fundamentally about. For this community of specialists, the stereotype for x_k (call it x_s) will include factors which go beyond everyday knowledge, and may incorporate a stereotype of x_e itself (with the *essence* of x_e a matter for scholarly debate).

Once this new stereotype (x_s) is established within the scientific community, it may begin to percolate through the rest of science into everyday use, eventually superseding y_s. Thus, a meaning change will occur – y will belong to x_k and not y_k. Consider gold as an example. The *essence* (and essence) of gold as the atomic number 79 is firmly established, with this property well known outside of scientific circles, possibly to the extent that it has become part of the stereotype. Thus, at present, the job of the scientist is to determine if a given lump of material has an atomic number of 79, and thus, is gold (i.e. to determine the extension of gold). However, current research might attempt to examine the *essence* of atomic number, which could be a matter for debate (controversy over theories for subatomic particles, for example). A major change in opinion at this level could have repercussions for the *essence* of gold, and potentially result in a change of meaning. It is perhaps here that the concept of hard and soft sciences originates. The 'hard' sciences have managed to alter everyday stereotypes of many natural kinds, even to the extent of incorporating their *essence* (gold has an atomic number of 79). This is something that the 'soft' sciences have failed to do as effectively. This model for reference fixing also perhaps gives an insight into the cause of the reductive tendencies of the hard sciences. Once an *essence* has been determined, then an important problem is solved (apart from the practical issue of determining extension). The next step is to query the assumptions underlying the definition of the *essence*, which necessarily results in a narrowing of the research focus.

Some important departures between my perspective and Putnam's own views are clear here, but these result from our different definitions of essence, which is absolute for Putnam, but is an *essence* that may change from my perspective. One consequence of my view is that different communities with varying degrees of expertise in a particular discipline are likely to use stereotypes of varying detail. Furthermore, some communities may be aware of the *essence* of a term and others may not. Thus, stereotype is not just conceived of at one level – that of the everyday language. The body of facts that a speaker needs to be in command of in order to engage in debate is a function of the expertise of the community that they are interacting with at that time. For a conversation between these individuals to be meaningful, it is necessary that there is a common core of agreed upon information – the community-specific stereotype.

Hopefully, the difference between my views and those of Kripke and Putnam are beginning to emerge. However, before I develop my argument further, it is important to address an issue of critical importance – do geomorphic natural kinds exist?

5.3 Are geomorphic objects natural kinds?

The simple answer to this question according to Wilkerson (1988) is a resounding 'no', with the implication that geomorphology is not scientific according to his rather strict definition of science. The reason that Wilkerson does not believe geomorphology can be a science is because the objects that we analyse have no essence (they also have no *essence*). Thus, these objects are not natural kinds and there can be no science of these objects because they are not defined in such a way that permits serious scientific generalization.

> *Geographers talk of cliffs, beaches, mountains, valleys, seas and volcanoes. Meteorologists talk of depressions, anti-cyclones, winds, thunderstorms, clouds and hurricanes. But none of these terms picks out a real essence, and none lends itself to scientific generalization. One and the same lump of material will count as a mountain in one environment, as a valley floor in another, and, in yet another, as part of the sea bed.*
>
> Wilkerson, 1988, 32

According to Rhoads and Thorn (1996, 121) this view '… may be difficult to refute given the current status of theory development in geomorphology'. One way to try and recover the scientific status of geomorphology from Wilkerson's critique is to argue over his definition of what constitutes science. However, a more effective form of argument accepts his definition and then tries to counter his objections upon their own terms (an immanent critique). I believe Wilkerson to be incorrect in his assessment of the status of geomorphic objects and I feel that in principle, the scientific status of geomorphology can be recovered from his argument. By implication, this suggests that my perspective on theory development within geomorphology may be different to that of Rhoads and Thorn.

The problem with Wilkerson's formulation is that the essence of a geomorphic object is taken to be 'the stuff from which it is made'. Not surprisingly given the quotation above, Wilkerson considers disciplines that directly study the materials that make up our planet (chemistry, for example) to be working with kinds that have an essence, and thus, to have a truly scientific perspective. Thus, there are 'hard' and 'soft' sciences, or perhaps even sciences and non-sciences. However, from the perspective of modern geomorphology, a definition of a geomorphic object that is based purely on constitutive materials and morphology is inadequate. The processes that create the landform are also of vital importance. Consider crag-and-tails (Hambrey, 1994), avalanche debris tails (Potter, 1969) and pebble clusters (Brayshaw, 1983). All of these landforms consist of a core stone of material with an extensive deposit on the lee side of the obstacle and sometimes a smaller proximal deposit. An obvious way to distinguish these landforms is by their scale. However, it is difficult to see any simple way in which scale can become a part of the essence for a kind. Firstly, part of the project of Kripke and Putnam was to define kinds that are things in themselves. It may be difficult to incorporate the relative scales of large and small within an absolute essence. Secondly, an absolute scaling is problematic due to the inherent variability of rock fragment sizes and process energy and these scales might not hold in all possible worlds. The size of the landform is likely to be some general function of the particular nature of the process operating at that location. Thirdly, landforms develop through time. Hence, without invoking a concept such as maturity (Melton, 1958) it is difficult to see how a specific scale can be defined.

This leaves process. If the essence of the form is tied to the formative process, then the difference in scale is implicit, given the physics of the process identified and the nature of the geotechnical material under consideration. On Twin Earth, the relative scaling of these three landforms may alter radically due to differences in the physics of glaciers, snow avalanches and water flow. However, if the definition couples the morphology to a process, the landform should be located in an environment that is a consistent with the whole suite of phenomena associated with that process (Twin Earth eskers and drumlins in the case of crag-and-tails, for example). Indeed, this type of definition is the most commonly used in contemporary geomorphology. For example, in *The Encyclopaedic Dictionary of Physical Geography* (Goudie et al., 1985), David Sugden defines a crag-and-tail as 'A streamlined landform comprising a rock obstruction with glacial deposits in its lee.' *The Oxford Companion to the Earth* (Hancock and Skinner, 2000) defines graben as 'an elongate block of crustal rocks that has been thrown down by normal faults dipping towards it on one or both sides', and yardangs as 'aerodynamically shaped desert landforms produced by the wind erosion of bedrock or sediments'.

Thus, has Wilkerson failed to understand the manner in which landforms are defined in contemporary geomorphology? Has he made the mistake of considering his stereotype to equate to the contemporary *essence*? Or is his point still legitimate because none of the essences of geomorphic objects or natural kinds offered here are anything like as specific as those for physical and chemical kinds, such as Putnam's example of 'water is H_2O'? A rather obvious shortcoming of Putnam's suggestion for the essence of water was pointed out by Kuhn (1990). Ice and steam are also H_2O but have a very different physical behaviour to water. This means that the essence for water can not merely be H_2O unless no scientific distinction was made between these states. Hence, the actual essence for water must be something closer to 'liquid H_2O'. The necessity to move beyond a description of the constituent elements, their relative abundance and their bonding, to describing the energy of these molecules, moves the essence of water from 'the stuff from which it is made' towards process. Thus, it is not illegitimate to introduce process-based components to the essence of geomorphic objects, which would suggest that geomorphic objects can be considered to be natural kinds, even if the determination of extension for these kinds is not trivial.

5.4 Geomorphic natural kinds and the scientific status of geomorphology

Hill, mountain, beach, brook and pebble are all words from ordinary language that refer to objects of concern to geomorphologists. Some of these words form part of the technical vocabulary of geomorphology, while others do not. Thus, geomorphology is in a situation where some landscape features with names in everyday language have a geomorphic *essence*, while others are general labels with no technical usage and no scientific essence. Therefore, the stereotype for 'hill' (as opposed to hillslope) is probably nearly identical for a geomorphologist and a non-expert, while that for 'beach' is likely rather different due to the geomorphologist's emphasis upon sediment accumulation over and above recreational usage. In the majority of cases, general terms for geomorphic phenomena are the same as those used in everyday language, and derive from this language. Similar stereotypes are held by a very wide community and are generally a reasonably accurate reflection of the geomorphic *essence*. However, there are also a great many terms used in geomorphology that do not form part of the vocabulary of everyday language. These may derive from the language of the region in which such landforms are common or were first identified, or the native language of the person who identified the landform. Examples of these types of terms include 'karst' and 'karren'. The fact that ordinary language is not familiar with the stereotype for a great many geomorphic terms suggests that the everyday language stereotype of geomorphology (if it exists at all) or physical geography, lacks a great deal.

This situation can be contrasted with that for physics and chemistry, where there has been a more prevalent tendency for the meaning of terms initially defined within the scientific discipline to become part of everyday language. Gravity, black holes, relativity and atoms are obvious examples of this phenomenon. Thus, there is a difference between geomorphology and physics/chemistry in the degree to which reference has been transmitted from the scientific communities to the everyday language community. Over the last fifty years, perhaps only 'plate tectonics' has had its reference percolate through the various earth science communities as far as the ordinary language community. Hence, in the UK where geomorphology is traditionally a part of Geography rather than Geology, the scientific status of geomorphology amongst a broader public will not have gained greatly from this process of reference transfer.

Introducing a natural kind term into common usage from a particular scientific discipline has the effect of raising the profile of that discipline. This must be so as soon as the everyday language user begins to inquire as to where this new term came from. For example, learning that the term 'black hole' originates in astronomy and physics informs the everyday language user that science has discovered something so new about the world around us that it needs a new word. Thus, ordinary language begins to become replete with terms from certain disciplines, raising the status of these at the relative expense of those for whom meaning transmission has been less successful.

A typical ordinary language stereotype for black hole might be, 'A term from astronomy for a collapsed star that absorbs everything, including light.' Of course, this stereotype does not fix the essence of black hole and is also not part of the stereotype for the physics community in general, or astronomers and astrophysicists in particular. For example, perhaps Hawking radiation would be an element of the stereotype for a more specialized community. Because the majority of objects analysed by geomorphologists (rivers, mountains, etc.) have names that come from everyday language, it is much more difficult for their scientific usage to displace everyday language meaning compared to a completely new term. Examples of this do exist (a whale is a mammal and not a fish these days) but these are relatively rare occurrences. Thus, disciplines such as geomorphology will always struggle to have a high public profile.

Because geomorphic kinds take their names from everyday language usage, there is also a danger that the scientific *essence* can be merely added to the everyday stereotype instead of replacing it. For example, the definition for 'cave' given by Goudie et al. (1985) is, 'A natural hole or fissure in a rock, large enough for a man to enter'. This is probably almost exactly the same as the non-scientific, everyday language stereotype for cave, but is not a scientific statement and does not denote a natural kind. This implies that Wilkerson's argument that geographic objects have no scientific essence and thus, that physical geography is non-scientific, might have some merit. Subsequently, Goudie et al. outline the processes that initiate cave formation in limestone rocks as solutional enlargement of joints in the phreatic zone, followed by a lowering of the water table and the initiation of vadose zone processes. Is this not a more effective position from which to define a scientific essence? If it is, then a cave is any passage through the rock that permits vadose zone processes to begin to operate, i.e. a cylindrical tunnel of diameter of 10 mm may constitute a cave. This may seem counter-intuitive, but this is only true relative to the everyday stereotype. Scientifically, it is much more effective to group landforms together on the basis of formative process rather than the rather *ad hoc* scale selection of the size of a human being. Thus, in order to fix an *essence* for some types of geomorphic objects, it may be necessary to reorganize the taxonomy somewhat. Instead of caves as natural openings in rock at a human scale, with subdivisions into the type of cave given by a process description (instead of a natural kind name), the formative process is used to define the name of the kind itself. This may require the introduction of some new terminology, but would result in a set of kinds with a scientifically driven *essence*.

Purely morphological definitions of landscape features also exist in geomorphology. In these cases it is again difficult to see how an appropriate scientific *essence* may be defined; a process-oriented definition permits much greater insight. It is not at all surprising given the emphasis within geomorphology of fieldwork and identification of landscape features that a morphologically driven taxonomy can result. However, once subsequent process-oriented research has determined the processes that form a particular landform, a reorganization of the initial taxonomy on the basis of process may permit a more scientific *essence* to be established. This may require some reorganization of taxonomies but, as has already been discussed, these changes of meaning can and do occur from time to time in other scientific disciplines.

Owing to the wonderful landforms that are produced in limestone environments, karst geomorphology is particularly susceptible to this morphological classification. Goudie et al. (1985) state that karren is a collective name for 'small limestone ridges and pool structures which have developed as a result of the solution of rock by running or standing water'. There are many types of karren, all differentiated by morphology. Sweeting (1972, 75) also adopts a morphological approach to karst.

However, it would appear that some karst scientists have appreciated the weakness of a morphologically driven approach. Bögli (1980) discusses a range of karren features and associates with each a different process-based description – an attempt has been made to move towards a more scientific determination of *essence*. However, this very much welcomed effort still suffers from the fact that the same terminology is used, which was originally morphologically oriented, leading to potential confusion, even between scientists, over the nature of the *essence*. A clean break with morphologically driven terminology might be more profitable. For example, Bögli (1980, 60) defines pseudokarren as features that 'develop on insoluble, mostly silicate rocks by means of weathering processes. They show a rounded type of rinnenkarren, less frequently the form of atypical solution pans.' Thus, the process-oriented explanation is used to explicitly recognize that pseudokarren are not true karren features, but the use of the suffix 'karren' still implies a link (driven by morphology) to more classical karren features. Pseudokarren are formed upon different rock types by different physical processes and thus, are nothing to do with karren at all. By failing to recognize this in our landform taxonomies, we continue to confuse morphological definitions with process-oriented *essences*.

Recently, there has perhaps been a non-explicit recognition of the issues touched upon in this paper within other parts of geomorphology. Olav Slaymaker, who proposed in the 1980s that geomorphology moved towards a research group model for research (Church et al., 1985), recently edited a collection of papers under the title of *Steepland Geomorphology* (Slaymaker, 1995). In his introduction to this volume, Slaymaker justifies his choice of 'Steepland' (a well-known term in New Zealand) and tries to delimit its usage. Steeplands are 'steep and tectonically active areas with unstable landscapes, where erosion, sediment production and sediment transport rates are highly variable and difficult to predict because of inadequate theory and models' (Slaymaker, 1995, 1). This definition is interesting from at least two standpoints:

1. It implicitly recognizes that *essence* is not static and can be renegotiated within the scientific community. Dramatic improvements in theory and models (a 'paradigm shift' for the steepland geomorphological community?) might mean that the latter part of the definition no longer held.

2. Steepland geomorphology incorporates mountain geomorphology as well as various other areas, including types of hillslope geomorphology.

Both mountain and hill are everyday language terms with well-established stereotypes. The introduction of a new term with a scientifically driven *essence* removes the confusion between everyday language stereotype and scientific usage. This gives the scientist a linguistic breathing space and removes barriers to reference transmission that spring up because of a presumed identity between scientific and common language stereotypes. Thus, with the movement towards process-oriented reference fixing and the replacement within geomorphology of common language words by new terms with no *a priori* stereotype, it is possible to begin to move towards a geomorphology that defines the nature of the features of our landscape for all linguistic communities. The linking of *essence* to process produces a classification of landforms that is scientific in the sense discussed by Wilkerson (1988) and the new terminology permits members of other linguistic communities to learn about our discipline without having first to unlearn their previously held stereotypes for the terms that we use. In this way, geomorphology can raise its status to that of a science that does help define the way in which many communities see the world around them. As it creates its own space in language, geomorphology can begin to define in its own way the manner in which science is perceived, something that at present is left mainly to physics and chemistry.

5.5 Scientific revolutions and changes in meaning

Kuhn (1962) argued that from time to time, the process of normal science goes through a crisis, the end of which is indicated by a paradigm shift and a change in the way we engage with the world (the living in different worlds argument). Rather important to this argument was the idea of a succession of these events, an idea disputed by Lakatos (1970), who allowed for individual research programmes to wane as others developed. (It is important to note that Lakatos also rejected Kuhn's relativism and tried to define rational criteria for the acceptance and rejection of different research programmes.) The view of Feyerabend (1975) was that scientists should always try and break out of the paradigm that they work under by coming up with outrageous hypotheses to challenge the dominant normal science hegemony. These three views are perhaps the best known within geomorphology as a whole.

One consequence of the view of natural kinds taken here is that not only can different paradigms exist at a given time, they can also be held by different groups of people. The notions of stereotype, *essence* and reference transmission permit changes in scientific thought (exemplified by the different worlds argument) to be conceptualized in a new way. The important idea here is that of a research community. A paradigm shift may occur within a particular research community and revolutionize research in that field, giving a totally new outlook on the way the world works. However, the speed and extent to which this moves through different scientific communities to cause a change of meaning at the level of everyday language is a function of several phenomena, including the ease of reference transmission (the nature of terminology, the strength of pre-existing stereotypes, the linguistic space of the discipline), the importance of the idea for different communities and the politics of the way in which an attempt is made to cause a change in meaning. Thus, at any given time, different communities hold a range of stereotypes associated with different ideas. A geomorphologist may have stereotypes of certain branches of physics that are identical to the everyday language community, a stereotype of certain areas of geology that is little better, while within their own discipline they may have caused a change in *essence* and extension that only a dozen people in the world are aware of. There is a fluid hierarchy of scientific revolutions with associated changes in meaning, not the single scale of change envisaged by Kuhn (1962) and Lakatos (1970). The nature of scientific change described here is much closer to the more recent views of Thomas Kuhn (1990) and justifies a more sophisticated investigation into scientific sociology. At present, the inertia of certain everyday language stereotypes for terms in geomorphology may well prevent a common language paradigm shift from ever occurring.

Acknowledgements

My interest in some of the issues discussed here was first awoken during tutorials with Barbara Kennedy and Heather Viles at Oxford in 1994. However, it was a talk by Brian Whalley in 1997 that encouraged me to look into this topic in greater detail.

References

Bauer, B.O. 1996: 'Geomorphology, geography and science'. In B.L. Rhoads and C.E. Thorn (eds.), *The Scientific Nature of Geomorphology*. John Wiley, Chichester.

Bögli, A. 1980: *Karst Hydrology and Speleology*. Springer-Verlag, Berlin.

Brayshaw, A.C. 1983: *Bed Microtopography and Bedload Transport in Coarse-Grained Alluvial Channels*. Unpublished PhD thesis, Birkbeck College, University of London.

Church, M., Gomez, B., Hickin, E.A. and Slaymaker, O. 1985: 'Geomorphological Sociology'. *Earth Surface Processes and Landforms*, 10, 539–540.

Dupré, J. 1981: 'Natural kinds and biological taxa'. *The Philosophical Review*, 90, 66–90.

Feyerabend, P.K. 1958: 'An attempt at a realistic interpretation of experience'. *Proceeding of the Aristotelian Society*, 58, 143–170.

Feyerabend, P.K. 1970: 'Consolations for the specialist'. In I. Lakatos and A. Musgrave, (eds.), *Criticism and the Growth of Knowledge*. Cambridge University Press, Cambridge, 197–230.

Feyerabend, P.K. 1975: *Against Method,* Third Edition. Verso, London.

Frodeman, R. 1995: 'Geological reasoning: Geology as an interpretive and historical science'. *Geological Society of America,* 107, 8, 960–968.

Goudie, A.S., Atkinson, B.W., Gregory, K.J., Simmons, I.G., Stoddart, D.R., Sugden, D. 1985: *The Encyclopaedic Dictionary of Physical Geography*. Blackwell, Oxford.

Hacking, I. 1990: 'Natural Kinds'. In R.B. Barrett and R.F. Gibson, *Perspectives on Quine*. Blackwell, Oxford.

Hacking, I. 1993: 'Working in a new world: The taxonomic solution'. In P. Horwich (ed.), *World Changes: Thomas Kuhn and the Nature of Science*. MIT Press, Cambridge, Massachusetts.

Hambrey, M. 1994: *Glacial Environments*. UCL Press, London.

Hancock, P.L. and Skinner, B.J. 2000: *The Oxford Companion to the Earth*. Oxford University Press, Oxford.

Horton, R.E. 1945: 'Erosional development of streams and their drainage basins: hydrophysical approach to quantitative morphology'. *Bulletin of the Geological Society of America*, 56, 275–370.

Kripke, S. 1972: 'Naming and necessity'. In D. Davidson and G. Harman (eds.), *Semantics of Natural Language*, Reidel, Dordrecht.

Kripke, S. 1980: *Naming and Necessity*. Blackwell, Oxford.

Kroon, F.W. 1985: 'Theoretical terms and the causal view of reference'. *Australasian Journal of Philosophy*, 63, 2, 143–166.

Kuhn, T.S. 1962: *The Structure of Scientific Revolutions,* Second edition. University of Chicago Press, Chicago.

Kuhn, T.S. 1990: 'Dubbing and redubbing: the vulnerability of rigid designation'. In C.W. Savage, (ed.), *Scientific Theories*. Minnesota Studies in the Philosophy of Science, volume 14, University of Minnesota Press.

Lakatos, I. 1970: 'Falsification and the methodology of scientific research programmes'. In I. Lakatos and A. Musgrave, (eds.), *Criticism and the Growth of Knowledge*. Cambridge University Press, Cambridge.

Locke, J. 1690: *An Essay Concerning Human Understanding*.

Melton, M.A. 1958: 'Geometric properties of mature drainage systems and their representation in an E4 phase space'. *Journal of Geology*, 66, 35–56.

Potter, N. 1969: 'Tree ring dating of snow avalanche tracks and the geomorphic activity of avalanches, Northern Absaroka Mountains, Wyoming'. *Geological Society of America,* Special Paper, 123, 141–165.

Putnam, H. 1975a: 'Is semantics possible?'. In H. Putnam *Mind, Language and Reality: Philosophical Papers, Volume 2*. Cambridge University Press, Cambridge.

Putnam, H. 1975b: 'Explanation and reference'. In H. Putnam *Mind, Language and Reality: Philosophical Papers, Volume 2*. Cambridge University Press, Cambridge.

Putnam, H. 1975c. 'The meaning of "meaning"'. In H. Putnam *Mind, Language and Reality: Philosophical Papers, Volume 2*. Cambridge University Press, Cambridge.

Rhoads, B.L. 1994: 'On being a "real" geomorphologist'. *Earth Surface Processes and Landforms*, 19, 269–272.

Rhoads, B.L. 1999: 'Beyond pragmatism: the value of philosophical discourse for physical geography'. *Annals of the Association of American Geographers*, 89, 4, 760–770.

Rhoads, B.L. and Thorn, C.E. 1996: 'Towards a philosophy of geomorphology'. In B.L. Rhoads and C.E. Thorn (eds.), *The Scientific Nature of Geomorphology*. John Wiley, Chichester.

Richards, K.S. 1990: '"Real" Geomorphology'. *Earth Surface Processes and Landforms*, 15, 195–197.

Simpson, G.G. 1963: 'Historical Science'. In C.C. Albritton (ed.), *The Fabric of Geology*. Freeman, Cooper and Co., Stanford, CA.

Slaymaker, O. 1995: 'Introduction'. In O. Slaymaker (ed.), *Steepland Geomorphology*, John Wiley, Chichester.

Sweeting, M.M. 1972: *Karst Landforms*. Macmillan, London.

Wilkerson, T.E. 1988: 'Natural Kinds'. *Philosophy*, 63, 29–42.

Wilkerson, T.E. 1993: 'Species, essences and the names of natural kinds'. *The Philosophical Quarterly*, 43, 170, 1–19.

Wittgenstein, L. 1958: *Philosophical Investigations*, Second edition. Blackwell, Oxford.

6

Putting the morphology back into fluvial geomorphology: the case of river meanders and tributary junctions

André Roy and Stuart Lane

6.1 Introduction

One of the common misconceptions about geomorphology is that, whilst other parts of geography have sought to define, evaluate and re-define their core philosophical and methodological approaches, it has been satisfied with the consequences of major upheaval in the 1950s and 1960s. The works contained in Rhoads and Thom (1996) emphazies that this is emphatically not the case. However, following on from the major changes in approaches to geomorphology in the 1950s and 1960s, it is probably fair to say that geomorphologists have been largely complacent with a broadly positivistic conception of the way in which the world works: theoretically informed empirical regularities. This approach was identified by Strahler (1952), who drew upon the works of Gilbert (1877) and Horton (1945) to argue that geomorphology needed a process-based approach to the explanation of landforms, as distinct from the Davisian view of landscape development that was dominant at the time. However, this paper identifies selected trends within fluvial geomorphology to argue that, rather than being ontologically uninspired, some fluvial geomorphologists have been innovative, albeit implicitly and perhaps unintentionally, in changing their view of the way that rivers behave. There has been the progressive growth of locally specific, intensive research into particular river channel reaches, as distinct from the earlier research, which was grounded in the search for empirical regularities or laws. This paper seeks to show that this changing view has strong parallels with other aspects of geography: a move away from the development of general models for the prediction of human behaviour towards a recognition of the complex nature of many individuals within society as a whole. This reflected growing concerns with the ability of research to deal with the activities of complex individuals through aggregate models of human behaviour. Whilst such models may have been effective in describing general patterns, doubt was expressed both over their ability to explain the fundamental reasons for those patterns and their tendency to exclude from analysis those individuals who did not fit the rule. In the same way, fluvial geomorphology is now increasingly grounded in the measurement and understanding of individual cases.

These changes raise difficult questions over the form and nature of generalization that is or may be achieved. This is in a discipline that, whilst being commonly based within geography, requires a strong and explicit association with the natural sciences, where the search for generalization through experimentation is a tradition. A contradictory situation seems to arise: the focus on individual reaches in particular field environments eliminates the possibility of simple replication and controlled experimentation, as what one finds is a product of where one studies and when. Thus, the elaboration of laws becomes difficult, as everywhere seems to be different. Those who study particular cases become criticized by those who view the subject from a different perspective, as knowing everything about something but nothing about anything.

This paper aims to show that central to the rationale behind this change in fluvial geomorphology is the recognition that morphology exerts an active and conditioning role over the processes that operate within the river. This does not preclude the existence of generalizations, but it does require that they be viewed in a different way. In the same way that Giddens (1981) describes how structures both constrain and enable human activity, so current research is concerned with the way in which particular types of morphology constrain and enable particular types and intensities of process, in a manner that, like society, changes in response to those processes. The recognition of morphology as a key influencing variable is not new (Richards, 1996; Church, 1996; Lane and Richards, 1997), but rarely has it been stated, particularly in terms that relate to the wider geographical community.

This paper, therefore, attempts to bring to the attention of the whole of geography the extent to which physical geography is not entirely complacent (Spedding, 1997 and Lane, 2001b). As such, it seeks to re-open the dialogue between physical and human geography at an ontological level rather than simply a methodological one. This paper is about how we view the world, and not the way we go about studying the world, although we argue that the two are implicitly connected. We approach this issue by showing how, from within the subject, this view has been challenged, to such a degree that the approaches to the study of river channels can now be viewed as different to those of 40 years ago, and then by considering the relationship between scale, process and form in understanding these changes and the implications of these changes for the discipline.

6.2 Putting form back into morphology

Perhaps one of the most surprising aspects of recent research in fluvial geomorphology is its concern with progressively more localized, and often smaller scales of enquiry. This has involved the growth of intensive study for finite time periods of particular reaches of river channels. This may be labelled as case-study research (Richards, 1996), and this section of the paper argues that a key reason why there has been this progressive increase in emphasis upon a case-study approach can partly be explained by the inadequacies of earlier approaches and the search for a deeper understanding of nature. This is explored with reference to both river channel meanders (following Richards,

1996) and river channel confluences. Whilst the choice of examples to support this shift is selective, and partly a product of the view that we seek to bring to the history of river channel research, they do illustrate that fluvial geomorphology has broadened its approach to understanding river channel behaviour quite considerably since the original positivist approaches of the 1950s and 1960s.

6.2.1 Meanders

Meander research during the 1950s and 1960s was strongly grounded in the search for general empirical regularities at the reach-scale (Leopold and Wolman, 1957). Once the basic form of the regularities was recognized, field and laboratory research sought to refine these models. In the process of doing this, there was the progressive incorporation of new variables to account for the observed scatter (Schumm, 1960 and Ferguson, 1973). The focus remained reach-scale generalizations, but these had to be increasingly tuned to fit the observed complexity of the real world. The late 1960s and 1970s began to see a move away from this. Rather than seeking to produce general empirical regularities at the reach-scale, there was a progressive emphasis upon the description of meander form within the reach-scale, as the basis of explaining meander bend generation. Initially, this took the form of descriptive models of meander bend regularity using sine curves (Ferguson, 1975), and with time this moved towards descriptive models of irregularity, with disturbances to regular curves introduced both through the structure of the sine curve descriptions and through periodic and stochastic disturbances of those descriptions. This reduction in scale was reinforced by the growth of within-channel process investigations.

Whilst the study of secondary flows in meander bends had a long history and important advances had been made in the field during the 1960s (Rosovskii, 1954), the 1970s saw some of the first major measurements of time-averaged secondary flow structures, using the newly available electromagnetic current meters. Thus, Thorne and Hey (1979) introduced the idea that in situations of sharp bend curvature, it was possible to get two helical circulation cells within a given section, with upwelling both at the outer edge of the bend and up the inner-edge point bar (Figure 6.1). Research sought to establish the presence and absence of this type of flow structure, in a range of bend environments, and this was coupled with investigation of the implications of curvature for processes like sediment transport (Bridge, 1977 and Bridge and Jarvis, 1982). A series of investigations into the geomorphology, sedimentology and hydrodynamic processes of particular bends (Hickin, 1974 and Hickin and Nanson, 1975 and Carson and Lapointe, 1983) was epitomized by the Dietrich studies (Dietrich et al., 1979; Dietrich and Smith, 1983, 1984 and Dietrich, 1987) of a single meander bend. The latter were important in identifying the importance of irregular bend morphologies in controlling the nature of resultant processes. Dietrich and Smith (1983) show that meander-bend research to date had been based upon a poor understanding of hydrodynamic theory, that the Rosovskii (1954) analysis of field data produced misleading results, and that central to meander bend processes was the role of topographically induced convective

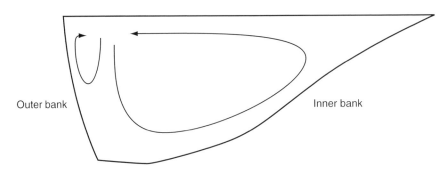

Figure 6.1 *A schematic of the hypothesised twin, back-to-back helical circulation identified by Thorne and Hey (1979). The larger circulation cell, located over a point bar, is associated with the effects of streamline curvature, resulting in water surface super elevation towards the outer bank of the channel and depression towards the inner bank. This results in pressure gradients leading to flow at the bed, or secondary circulation towards the inner bank. The outer bank circulation was hypothesized as being associated with the effects of bank and vegetative roughness.*

accelerations. This became the subject of some debate, but was ultimately reconciled in the acknowledgement that field meanders retained special morphological features, which was why field-based conclusions attesting to the importance of convective accelerations differed from laboratory ones.

6.2.2 Confluences

It is widely recognized that river confluences are a key component of the structure of drainage networks and that they play a critical role in the routing of flow and sediment transport throughout the fluvial system. In recent years the study of confluences has become an important research program in geomorphology that has evolved from the investigation of regularities in planform geometry of river junctions to the detailed measurement of physical processes at a few selected sites.

Very early, it had been noted that the geometry of river confluences exhibits some regularity throughout the drainage network. Hutton (Playfair, 1802) had made two critical observations: river channels merge at the same height at confluences, a feature described as being concordant, and smaller tributaries enter the main channel at right angles whilst rivers of similar size merged at more acute angles. These observations prompted Horton (1932, 1945) to propose a model for explaining and predicting confluence angles. His rationale was based on the fact that water flows downhill following a path of maximum slope. This implies that small streams flowing on a steep hillside will tend to enter the main stream at a right angle. As slope angle decreases so does the angle of entry of the tributary into the main channel. Horton's model applied well to smaller streams entering a large one, but did not yield satisfactory predictions for equal size streams merging at a confluence. This has led

Howard (1971) to develop an optimal model for stream junction angles. The model uses hydraulic geometry relationships in conjunction with a minimization procedure for power losses that optimizes the angular geometry of the junction by adjusting the length of the channels. This model was later generalized by Roy (1983a), who proposed several other optimality criteria. These optimal solutions reproduced the average regularities observed in the drainage network, but did not predict well the geometry of individual junctions (Roy, 1983b). Despite this shortcoming, optimal angular geometry models are still used in landscape modelling strategies (Rodrigez-Iturbe and Rinaldo, 1997).

One difficulty in the assessment of the performance of the models is that their failure to predict angles at individual junctions either may reflect problems in the optimization approach or inaccuracies in the use of hydraulic geometry relationships, or both. Roy and Woldenberg (1986) have shown that hydraulic geometry exponents are extremely variable at individual junctions and that a wide range of different morphological adjustments was a characteristic of confluent channels. Hydraulic processes at confluences appeared to be complex, and this seems to necessitate a more intensive investigation of hydrodynamic behaviour within specific confluences. This focus on physical processes would hopefully exhibit characteristics suitable for further generalization. Studies of confluence dynamics appear to be required.

Early work on the dynamics of river confluences was conducted largely by engineers. For instance, Taylor (1944) and Webber and Greated (1966) examined the changes in water surface height at confluences using an approach based on momentum. As expected, their work showed a sensitivity of the adjustments in water depth upstream and downstream of the confluence to junction angle. As angle increases, the effects of merging the flows were more severe. By combining junction angle with the discharge ratio between the two confluent channels, they obtained excellent predictions of the water depths simulated in a laboratory flume. Despite the importance of these theoretical and experimental advances, the approach did not provide geomorphologists with the necessary information to understand changes in channel morphology at river confluences.

In a pioneering study, Mosley (1976) used the knowledge that both junction angle and discharge ratio were key variables in the dynamics of river confluences to generate a series of flume experiments where he examined the relationships between fluid flow and bed morphology at confluences. His work has shown that a deep scour was created at river confluences, something that has been noted in both small confluences (Ashmore et al., 1992; Biron et al., 1993b; Rhoads and Kenworthy, 1995; and Rhoads, 1996) and very large confluences (Best and Ashworth, 1997). Furthermore, scour depth increases with the ratio of discharges between the tributary channel and the main channel and with the junction angle. Mosley also observed the water surface and the flow structure using injected dye. He reported the presence of back-to-back helicoidal cells that converge at the centre of the confluence to plunge towards the bed and roll back towards the banks of the confluence (Figure 6.2). These cells were the mechanisms that produced the scour region in the confluence and controlled the pathways of sediment transport. Mosley and others have reported similar processes in confluences in both dendritic drainage networks (Rhoads and Kenworthy, 1995; Rhoads, 1996; Rhoads and Kenworthy, 1998; and Bradbrook et al., 2000) and braided rivers (Ashmore, 1982; and Ashmore et al., 1992). Of most importance is the work of Rhoads (Rhoads and Kenworthy, 1995; and Rhoads, 1996), which demonstrates the interaction between momentum ratio, a more physically-meaningful measure of the discharge ratio, and confluence morphology, in a field site where the reach-scale planform, and hence junction angle, is relatively stable. This work also described the downstream evolution of back-to-back helicoidal cells. Bradbrook et al. (2000) used numerical modelling to show how this evolution was related to the combined effects of the radii of curvature and flow momentums in the tributary channels.

Thus, this demonstrates how large-scale confluence morphology, which is relatively stable, interacts with the more variable series of hydrographs delivered by each tributary to the confluence, to determine the zone of maximum downwelling and the position of the shear layer, and hence the migration of the scour hole in response to individual flood events.

The findings of Mosley were followed by more detailed and systematic work by Best (1985, 1986, 1987, 1988) on asymmetrical junctions. Best established clear relationships between scour depth and both the angle of the confluence and the discharge ratio. He also showed that one of the main controls of bed morphology is the development of a separation zone at the downstream corner where the tributary channel enters the main channel (Best and Reid, 1984). The size and strength of the separation zone increase with angle and with the discharge ratio. This controls the location and orientation of the scour zone and consequently of bedload sediment transport pathways, which follow the edges of the scour region. Best observed the presence of helical motion within the confluence, but also discussed the relative role of these cells with respect to the shear layer developed in between the confluent flows and to the flow separation which occurs at the edge of the scour region (Figure 6.3). Although more complex than previous models, Best's proposition seems to provide a comprehensive explanation of the dynamics and morphology of confluences.

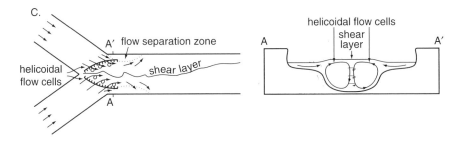

Figure 6.2 *The development of helicoidal flow cells at a symmetrical confluence according to Mosley (1976) (modified by Best, 1986).*

This view was changed through a series of independent laboratory and field observations. Ashmore and Parker (1983) looked at the relationship between scour depth and the two key variables used in laboratory work. The relationships that they obtained showed considerable scatter. There were no satisfactory predictions of scour depth from either the junction angle or the discharge ratio. Kennedy (1984) examined several confluences in order to assess if they were concordant, as expected from Hutton and Playfair's speculations. She reported that concordant confluences were rare and that most river junctions had a channel bed that is higher than the other. Roy and De Serres (1989) observed very complex flow dynamics at a small confluence with discordant beds. At this small confluence, flow mixing was extremely intense, and upwelling of the main channel flow was clearly visible in the separation zone downstream from the tributary. This fortuitous observation, along with the realization that current models could not account for the observed variability in geometry and morphology at confluences, led to a new series of research questions. These focused mainly on the role of bed discordance on the dynamics of confluences. Considering that bed discordance is probably the rule rather than the exception in nature, it is interesting to note that all previous work in the laboratory had always used concordant junctions. Therefore, the models that were developed from this work were only applicable to a small subset of confluences.

Using a very simple confluence model, Best and Roy (1991) have shown that bed discordance between two parallel channels entails major effects as the flow separation zone, developed in the lee of the step, interacts with the mixing layer between the merging flows (Figure 6.4). This interaction produces a distortion of the mixing layer and enhances mixing of the two fluids. This simple experiment gave an impetus to probe further into the effects of bed discordance at confluences.

Through a field survey, Biron et al. (1993b) showed that the morphology of confluences with bed discordance was not dominated by a scour zone but rather by the development of a tributary mouth bar that prograded into the confluence. Its extension may vary in

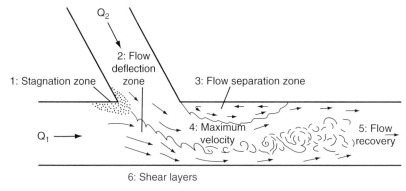

Figure 6.3 *Flow model at an asymmetrical river confluence depicting six distinct zones as proposed by Best (1987).*

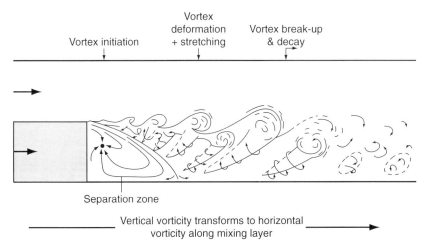

Figure 6.4 *Schematic view from above of the flow structure at the confluence of two parallel channels of unequal depth. The channel at the top is deeper than that at the bottom. Note the distortion of the shear layer as its base is gradually entrained on the side of the shallower channel where fluid upwelling occurs (from Best and Roy, 1991).*

response to changing flow conditions. The edge of the bar was often controlled by the position of the shear layer between the confluent flows, which was the dominant dynamical feature of these confluences (Biron et al., 1993b; Leclair and Roy, 1997). The shear layer was characterized by very high turbulence intensities that are more than 10 times the intensities observed in the channels upstream from the confluence (Biron et al., 1993a; and De Serres et al., 1999). These results from the field were corroborated by studies in the laboratory (Biron et al., 1996a, b) and by recent numerical modelling of discordant systems (Bradbrook et al. 1998, 2001). The effects of bed discordance are also of critical importance in assessing the distance of flow mixing downstream of confluences. Gaudet and Roy (1995) reported that flow mixing at confluences with discordant beds is 10 times shorter than those previously reported in the literature. The rapid flow mixing at discordant confluences is also corroborated by numerical modelling (Bradbrook et al., 1998). Despite these rather general statements on the behavior of discordant confluences, the results of this research have led to a complex picture of the dynamics of confluences as illustrated by the model proposed by De Serres et al. (1999) shown in Figure 6.5.

Research on river meanders and confluences provides good examples of what has happened in fluvial geomorphology. The development of knowledge has evolved in a direction where increasing complexity has been addressed. This has profound consequences on what appeared to be general models either to predict the geometry and morphology of fluvial features or to explain the physical processes that are characteristics of these sites. The shortcomings of the earlier models become rapidly evident when comparisons were made with the natural environment showing that there exists considerable unexplained variability. One of the major contributions of this research is that bed morphology is not just the product of the physical processes occurring at a confluence. Bed morphology exerts a control on the processes and it generates a hydraulic environment where processes are initiated that will feed back onto the form. This concept is not new but its importance is illustrated here in a very convincing manner. This has a corollary on the study of rivers: it will be extremely difficult to develop a general model of meanders or confluences. The initial conditions (i.e. the initial form) of each realization have to be taken into account in the development of its own dynamics. This implies that each study has its unique component that is the product of the history of physical processes.

6.3 Ontological evaluation

What this evidence has sought to do is to show how the dominant approach to the study of two typical fluvial landforms has changed through time. This change has the following typical: (1) a progressively greater recognition of the interdependence that can occur between form and process, which has resulted in the rediscovery of morphology as an influence upon process (e.g. Ashworth and Ferguson, 1986), rather than a passive response to process; (2) an associated reduction in the spatial and temporal scales of enquiry, with the study of particular river reaches or case studies; and (3) the need to develop new methods and new approaches, notably for the measurement of morphology (Lane et al., 1994 and Lane, 2001) and of flow processes (Roy et al., 1996; Lane et al., 1998; and Roy and Buffin-Bélanger, 2001) at high resolution over small spatial scales.

Making sense of this evolution is possible in two different ways. The first is ontological: it reflects the growing emphasis upon the mutual interaction between form and process in the understanding of geomorphological systems. The second is methodological and follows from issues that arise from this changing ontology, and is addressed in the next section.

Central to understanding the issue of form – process interaction, and the rediscovery of morphology, are questions of scale. As Schumm and Lichty (1965) observed, progressive increases in spatial and temporal scale change the status that we accord to form and process, between the classes dependent, independent and irrelevant. Almost all studies of meanders and confluences, as they have become more concerned with mutual interaction between form and process, have had to collapse the spatial and temporal scales of enquiry with the study of ever smaller bits of more manageable rivers. Whilst this reductionism does not necessarily require a reduction in spatial and temporal scale, in practice it normally always involves one. This provides us with a number of problems.

The first follows from Schumm and Lichty (1965). If the collapse of temporal and spatial scale has resulted in a scale of focus where form and process are mutually dependent, then questions emerge as to whether this mutual interaction applies at scales other than those which are the subject of this scale of enquiry. In essence, the emphasis upon feedback between form/morphology and process is simply a product of the study of ever smaller scales of enquiry. Second, and following from this, is the possibility that we have degenerated into detailology, in which this intensive research into the interaction between form and process tells us very little about the general behaviour of the system that we are interested in, whether over longer time scales or large space scales. This is an acute problem because of the growing concern within society about the longer-term impacts of environmental change, as well as the need for more holistic management of river systems based around the catchment as a natural geomorphological unit. Third, there is the question of big rivers. With notable exceptions (Best and Ashworth, 1997 and the International Association of Geomorphologists Commission on Large Rivers), the study of large river behaviour has been ignored. This may be because these rivers are not amenable to the dual form and process emphasis that comes with a form–process paradigm: measuring the morphology of big rivers is difficult (Lane, 2001a and Westaway et al., in press); measuring processes operating within them, especially during extreme events, is almost impossible.

The response to this can be both ontological and methodological. This progressive reduction in the spatial (and temporal) scales of enquiry could be viewed as being a product of finding that our existing models or views of the world are not quite commensurate with specific empirical experiences. The generalizations that we associate with the positivist analyses of the 1960s and 1970s are not necessarily incorrect; rather they are a manifestation of an analytical approach that does not recognize the importance of context as an influence upon system behaviour. The fact that there is inevitably scatter associated with empirical generalizations is an indication that other processes may be operating. When we see scatter, we are implicitly reductionist in our approach: in search for explanation, one progressively seeks a greater depth of understanding, especially of those things that do not fit. This does not necessarily involve a reduction in the spatial or temporal scales of enquiry, although in this case it does: from the basin scale through the reach-scale to the individual meander bend or river channel confluence. This change in viewpoint is not just reductionist, however. It has resulted in a very different conceptualization of the way that the world works, suggesting that we are not simply considering a scale change. Under the old view, meanders were seen as equilibrium channels, adjusted to what could be called a catchment-scale morphological regime. This adjustment manifests itself as some sort of statistically-defined empirical regularity which, following Strahler's (1952) argument, should lead to predictive and overarching laws. The progressive reduction in scale requires a new view of the way in which particularly river channel features work. An alternative view of rivers could be a postmodern one: every meander bend or confluence is unique and the creation of any sort of generalization is to exclude the very essence of the feature that is of interest. However, it can be argued that one set of overarching laws has been replaced with a different set, which allows a much stronger sensitivity to differences in morphology: form–process interactions in a river channel are conceived of as being associated with structure, mechanisms and events.

At the fluvial geomorphologist's scale of enquiry, classic Newtonian physics does perfectly well (Harrison and Dunham, 1998). Thus, we are given structures (the basic conservation laws of mass and momentum) that result in a set of mechanisms. These mechanisms emerge at the level of a particular river channel feature, and are defined by both overarching structures (e.g. conservation of mass, conservation of momentum) and the specific characteristics (e.g. morphology) of the features under investigation. A good example of a mechanism is secondary circulation in an irregular meander bend, where the irregular topography (an 'event') results in a change in the momentum balance (which is based upon momentum conservation or a 'structure') to produce secondary circulation, with outward directed flow across the point bar (a 'mechanism'). These mechanisms operate in a way that is defined by the morphology of the river channel feature; they, in turn, change the morphology of the river, and so the importance and behaviour of the mechanisms themselves are changed. This event–mechanism or form–process feedback gives the river a dynamic or evolutionary characteristic, which depends upon both the nature of the form–process interactions, and the role played by events that are exogeneous to the system. The result is that whilst it may be possible to produce reach-scale empirical generalizations, understanding of the behaviour of a particular section of river needs recourse to fundamentally different sorts of generalization: there *is* a shift from the general to the particular, but the result is a new sort of generality that has local manifestations that themselves are locally driven. In essence, time matters in the production of histories of particular places. There is a new regional (physical) geography, involving the study of particular places, but also involving a method of study and an ontological view of the world that is very different to the regional physical geography of W.M. Davis.

In addition to the idea that place becomes a crucial theme in physical geography, it is also important to recognize the potential for coupling across scales (Paola, 1996 and Lane and Richards, 1997). Whilst it might be a convenient research assumption that scale can be delimited into that which is of interest and that which is of a smaller scale and which can be either ignored or dealt with through simple parameterizations, but this assumption is problematic if what happens at smaller space and time scales does indeed matter for larger spatial scales or longer time periods. This can be illustrated with reference to river channel confluences. As noted above, time-averaged circulation (Mosley, 1976; Ashmore et al., 1992; McLelland et al., 1996; Rhoads and Kenworthy, 1995, 1998; Richardson and Thorne, 1998; and Bradbrook et al., 2001), such as back-to-back helicoidal structures, has been widely attributed as responsible for the generation of scour at confluences, associated with strong downwelling which acts as a momentum pump (Paola, 1996) and transfers higher momentum fluid towards the bed. This time-averaged circulation has been identified in the field and the flume by a series of point measurements of two or three flow components. Two problems arise. First, in the absence of a scour-hole, the strength of this downwelling is strongly reduced (Bradbrook et al., 2000). Thus, there is a basic problem of arguing that the downwelling is driven by a momentum pump and results in a scour-hole forming, when the presence of the scour-hole is the reason for the momentum pump being present. Second, some research has identified highly turbulent flow structures in confluences (Biron et al., 1993a) that do not always result in such time-averaged helical circulation (Figure 6.5). Furthermore, particularly in field environments, it is impossible to measure at more than a few points simultaneously (De Serres et al., 1999; McLelland et al., 1996; and Roy et al., 1999), and often only one point may be measured at a time (Rhoads and Kenworthy, 1995, 1998), and not always in three dimensions simultaneously (De Serres et al., 1999; McLelland et al., 1996; Rhoads and Kenworthy, 1995, 1998; and Roy et al., 1999). Recent measurement approaches have addressed this successfully (Rhoads and Sukhodolov, 2001; and Sukhodolov and Rhoads, 2001), and developments in numerical modelling have allowed assessment of this problem. Lane et al. (2000) use a three-dimensional model, coupled with an unsteady turbulence treatment (large eddy simulation) to show that the time-averaged flow model predictions may show very little resemblance to the series of flow field predictions from which they are derived. The trajectory of single particles is very different to the path defined by spatial pattern of vectors of average velocity. This observation is important, as large velocity fluctuations around an average state have been shown to be very important for sediment transport (Drake et al., 1988;

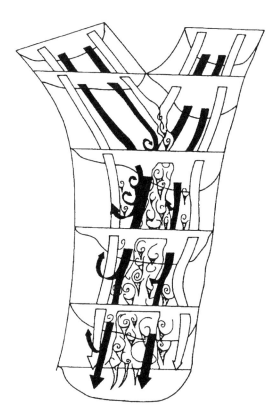

Figure 6.5 *Schematic flow representation of the observations made at the Bayonne-Berthier, a confluence with discordant beds. Note the complex interactions of between the confluent flow and the presence of upwelling of flow from the deeper channel on the side of the tributary channel (from De Serres et al., 1999).*

and Best, 1993). Shear between confluent flows will set up vortex structures, which could result in significant transfer of momentum towards the confluence bed, and produce periodically high drag and lift forces, which cause particle movement (Boyer, 1996).

This means that we must develop the basic momentum pump theory as the downwelling over the scour-hole exists only for a very small proportion of the time: consideration of coupled velocity observations and particle trajectory studies during scour suggests that scour is actually associated with high-frequency velocity fluctuations induced by shear between the tributary flows. There is a momentum pump, but it is driven by turbulence rather than time-averaged circulation, and confluence scour may be explained as being associated with the intermittent entrainment of particles due to shear-driven-turbulence. The multi-directional nature of flow fluctuations at the point of vortex attachment diminishes the traditional role of particle arrangement in reducing entrainment, so explaining why confluences commonly have scour-holes, but it is still necessary to understand the timing, magnitude and direction of a particular flow event before it is possible to say exactly when a particular particle will move. As particles around a given particle organize themselves in response to flow events, so the ability of that particle to move also changes. Thus, *when* a particle moves depends upon where it is, where it has been, how it relates to other particles around it, and how it conditions the flow field around it in a manner that is more or less conducive to movement. Deciding when a particular particle moves (or does not move) depends upon both the size and magnitude of impinging flow events (which may themselves be created by the position of the particle with respect to the flow) and the sequence of previous flow events, as these will determine the readiness of the particle to move. Small events may be very important if the time and place is right. There is a strong bi-directional feedback between form (in this case the morphology of a particular particle and its immediate neighbours) and process, such that there is a distinct history to particle movement within a particular place. Thus, the larger-scale scour-hole is a dynamic product of a series of smaller-scale events.

By collapsing the time and space scales of enquiry, we have a different model of scour-hole development, one that is grounded in turbulence as opposed to time-averaged flow fields. The central question here is whether or not the emergent property (the scour-hole) can be explained without recourse to these smaller-scale events. In some senses, it can. As noted above, Ashmore and Parker (1983) were able to develop a general model for scour-hole development, based upon junction angle and discharge ratio, both of which interact to control the strength of time-averaged downwelling (Bradbrook et al., 2001). However, the large amount of scatter around such a relationship suggests that the large-scale properties of the flow cannot explain observed scour characteristics. Implicitly, what is going on at the small scale does not always emerge at the larger scale in the same way as our representations of larger-scale explanatory variables might suggest.

This view challenges our traditional view of scale dependence and independence. It is not simply that changing our space and timescales of enquiry changes our focus of enquiry, but processes couple across scales. What happens at the small scale results in processes at that scale that lead to the emergence of larger scale features. This emergence raises intriguing methodological questions, notably whether or not we need to know the properties of small-scale fluctuations in order to predict large-scale system emergence.

6.4 Methodological evaluation

In considering the ontological status of a form-process approach, a number of crucial methodological perspectives must be developed. The previous section responded to accusations that these new approaches lead to little more than reductionist detailology by arguing that place and scale are more subtle influences on system behaviour than we have traditionally assumed. In making this argument, we glossed over a set of methodological issues, and these are returned to now. In methodological terms, associated with the recognition of a form–process emphasis has been a shift in the way in which we define the nature of a geomorphological experiment, not towards the elimination of extensive research in favour of case-study research, but more to recognize: (1) the different sorts of information that different types of experiment may provide; and that (2) the intensive–extensive dichotomy is a false one as almost all experiments make use of a mix of both approaches.

It is worth evaluating these in relation to the conventional role attributed to an experiment. Harré (1983) identifies three broad reasons for conducting an experiment: (A) as formal aspects of method; (B) in the development of the content of a theory; and (C) in the development of technique. Each of these has a series of sub-activities (Table 6.1).

Table 6.2 takes these classifications and applies them to the case of river channel research, with reference to both extensive and intensive approaches. This is a useful exercise as it demonstrates both the complimentarity of these two approaches and the role that case-study research can play in fluvial geomorphology. This demonstrates the essential role of the case-study approach: it can provide information on the hidden mechanisms behind a known effect (B1); it can provide proof that particular ideas apply in particular cases (B2); and it has the important characteristic of challenging basic assumptions that we make about apparently simple phenomena (B3), and which we may use as the constituent rules for defining, understanding, explaining and predicting a phenomenon. These are roles that extensive approaches find difficult to fulfil, and which demonstrate the complementarity of a case-study approach: there are certain things, notably inductive law-building and the identification of generality, that only an extensive approach can effectively achieve. Thus, in a discipline like fluvial geomorphology that may be 'maturing', the change in emphasis, with a reduction to smaller spatial and temporal scales, may reflect a growing emphasis upon the case study as a means for refining, perhaps eventually re-defining, a set of well-established concepts through giving them greater explanatory depth.

Table 6.1 *The Harré (1983) classification of the reasons for undertaking an experiment.*

A.	*As formal aspects of method...*
A1	...to explore the characteristics of a naturally occurring process
A2	...to decide between rival hypotheses
A3	...to find the form of a law inductively
A4	...as models to simulate an otherwise unresearchable process
A5	...to exploit an accidental occurrence
A6	...to provide negative or null results
B.	*In the development of the content of a theory...*
B1	...through finding the hidden mechanism of a known effect
B2	...by providing existence proofs
B3	...through the decomposition of an apparently simple phenomena
B4	...through demonstration of underlying unity within apparent variety
C.	*In the development of technique...*
C1	...by developing accuracy and care in manipulation
C2	...by demonstrating the power and versatility of apparatus

Thus, as a result of a more intensive approach, there has had to be a commensurate development in the technologies required to yield the required data. Much of this has been concerned with yielding high-resolution data from new current meters (e.g. the electromagnetic current meter and acoustic Doppler velocimeter) and from three-dimensional numerical modelling. This reflects the natural structure of any scientific discipline, in which the publication of new research requires findings that are both original and significant.

Conclusion

Through two cases, we have looked at the recent advancement of knowledge in fluvial geomorphology in order to understand how the approach in this science has changed. Although the analysis presented here is by no means exhaustive, it nevertheless points to the emergence of a new way of doing and thinking the science. This way reconciles the framework of looking for regularities and, therefore, for laws in the landscape, with the realization that there exists tremendous complexity and variability in forms. The work on confluences, for instance, exemplifies the progressive discovery of controlling factors of their behaviour to a point where it becomes virtually impossible to develop a predictive model for their morphology and dynamics. These deficiencies of predictive models become even more obvious when local contingencies are added to the picture. A widespread recognition of complexity is replacing the simplicity that is often sought. We suggest that these conclusions may also be valid in other fields in physical geography and that the contemporary meaning of geomorphology and perhaps of the whole of physical geography is now marked by a move from the search for predictive relationships

Table 6.2 Application of the Harré (1983) classification of the reasons for undertaking an experiment to rivers research.

As formal aspects of method...		A case-study approach	An extensive approach
A1	...to explore the characteristics of a naturally occurring process	✔ The hydrodynamic behaviour of a specific river channel confluence	✔ The relationship between confluence junction angle and scour-hole depth
A2	...to decide between rival hypotheses	✔ A set of possible different explanations of what controls scour-hole depth evaluated at a specific case (e.g. the role of time-averaged momentum transfer to the bed versus turbulence transfers of momentum). Note the experiment here defines which hypothesis is relevant to the case being studied and not cases in general	✔ The use of statistical methods to test hypotheses based upon significant difference or association between relevant parameters
A3	...to find the form of a law inductively	✘ Difficult to find a law from a single case study unless there is sufficient time for that case study to exhibit a range of characteristics of the parameters of interest (e.g. long-term observation of changing junction angle in relation to scour-hole depth at the same confluence	✔ Key aspect of an extensive analysis in which laws are generated from pattern in controlling variables (e.g. as junction angle increases, scour-hole depth increases)
A4	...as models to simulate an otherwise unresearchable process	✘ Fieldwork implies it is researchable	✘ Fieldwork implies it is researchable
A5	...to exploit an accidental occurrence	✔ Common as an accidental occurrence is, an exception and exceptions from a rule allow us to develop better rules.	✘ Exceptions to a statistical generalisation are typically ignored in an extensive analysis except where they cause research to adopt a more intensive approach, perhaps because the exception suggests that the empirically informed rule is not quite right (Roy and Woldenburg, 1986)

A6	...to provide negative or null results	✓ It is possible to falsify from a single exception, and this is important as it can lead to further research into why it is a null result. Note the limitation here that the exception must truly be an exception	✓ This can be more difficult as there are always exceptions to rules, and these will be typically averaged out in the analysis

In the development of the content of a theory...

B1	...through finding the hidden mechanism of a known effect	✓ This is a classic use of a case-study approach: reductionist in the sense that an observed phenomenon (e.g. the junction-angle/scour hole relationship) is explained in terms of measured flow and sediment transport processes through time in a particular place	✗ Unusual, as often it can be difficult to see the mechanisms behind a dataset, although mechanisms may be exhibited by a dataset
B2	...by providing existence proofs	✓ Demonstrates that something exists	✗ May demonstrate relationships but not phenomena
B3	...through the decomposition of an apparently simple phenomena	✓ Another classic reason for a case-study approach. Note that this has an important role in practical application as the identification that the assumptions associated with a phenomena don't hold causes us to challenge the way that the phenomena is treated itself	✗ Often lack detailed knowledge needed to do this
B4	...through demonstration of underlying unity within apparent variety	✗ This is difficult as there is always uncertainty in the generality of a finding based upon a particular case	✓ Tends to lead to identification of generality

that would have a high degree of generality towards the development of locally driven concepts and models. These models are based on a thorough understanding of the physical processes involved and on evidence derived from sound experimental designs applied to local and specific cases. Although there exist some useful predictive relationships, geomorphology does not seem to be in a law-seeking mode, but rather is in a learning mode by applying laws, in particular, those of physics and chemistry, to particular and often unique cases. As a consequence, general statements on the behaviour of a meander or of a confluence offer little insight into the understanding of specific cases. Yet, knowledge obtained from specific cases may find a high degree of generality.

The role of form in geomorphology is also changing as it is not simply what we hope to understand and predict but also a controlling factor in the evolution of geomorphic systems. Form influences process that will result in the modification of the form. The initial and boundary conditions of a morphologic system include as a primary input the form itself, which controls to a large extent the intensity of the processes. In a river, form does influence its own flow field, which in turn maintains the form. A clear example of this is the role of bed discordance in the dynamics of river confluences. Experimental work, both in the laboratory and in the field, suggests that bed discordance controls turbulence intensity and the distribution of bed shear stress and, thus, the resulting bed morphology. This is confirmed by numerical modelling. Even more important is the fact that the initial bed morphology is also the result of its past history. The system is on a trajectory that needs to be taken into account in assessing the response of the river morphology to changes in the external controls, such as the hydrological regime or the sediment supply. The evolution of the system is not only form-sensitive but also dependent upon its history (Lane and Richards, 1997). The consequence of this is that both spatial and temporal contingencies have to be known in order to make useful predictions of the behaviour of the system. The integration of time and space is a critical issue that has been addressed in all fields of geography (Massey, 1999).

The recognition of the complexity and the uniqueness of most cases that we study is not a dead end for research and its quest for understanding nature. It does not indicate a return to a strictly descriptive way of doing science without any hope of looking for generalization. It does emphasize, however, the limitations of a reductionist and simplistic view of nature and signals that generalization must come from another approach. The fact is that we now have the tools to address effectively these issues and to test the effects of historical scenarios and initial conditions on the morphology of river systems. An approach is emerging from the interactions between the experimental side of the discipline, seeking to establish and quantify the presence of effects from potentially controlling factors, and the numerical modelling side, which can extend the knowledge obtained through these experiments by simulating scenarios of form sensitivity and history. In fluvial geomorphology, computational fluid dynamics can clearly be used to provide new hypotheses to be tested experimentally. This approach will lead to the building of coherent research programs that will include a wide variety of scientific activities that will benefit from interactions and exchanges. The future lies in this cross-breeding of approaches and in the recognition of their individual as well as collective benefits.

References

Ashmore, P.E. 1982: 'Laboratory modelling of gravel braided stream morphology'. *Earth Surface Processes and Landforms,* 7, 201–225.

Ashmore, P.E., Ferguson, R.I., Prestagaard, K.L., Ashworth P.J. and Paola, C. 1992: 'Secondary flow in anabranch confluences of a braided, gravel-bed stream'. *Earth Surface Processes and Landforms,* 17, 299–312.

Ashmore, P.E. and Parker, G. 1983: 'Confluence scour in coarse braided streams'. *Water Resources Research,* 19, 392–402.

Ashworth, P.J. and Ferguson, R.I. 1986: 'Interrelationships of channel processes, changes and sediments in a proglacial braided gravel-bed stream'. *Geografiska Annaler,* 68A, 361–371.

Bathurst, J.C., Thorne, C.R. and Hey, R.D. 1977: 'Direct measurement of secondary currents in river bends'. *Nature,* 269, 5628, 504–506.

Best, J.L. 1985: 'Flow Dynamics and Sediment Transport at River Channel Confluences'. Unpublished PhD dissertation, Birkbeck College, University of London.

Best, J.L. 1986: 'The morphology of river channel confluences'. *Progress in Physical Geography,* 10, 156–174.

Best, J.L. 1987: 'Flow dynamics at river channel confluences: implications for sediment transport and bed morphology'. In F.G. Ethridge, R.M. Fores and M.D. Harvey, (eds.), *Recent Developments in Fluvial Sedimentology,* SEPM Special Publication 39, 27–35.

Best, J.L. 1988: 'Sediment transport and bed morphology at river channel confluences'. *Sedimentology,* 35, 481–498.

Best, J.L. 1993: 'On the interactions between turbulent flow structure, sediment transport and bedform development: some considerations from recent experimental research'. In N.J. Clifford, J.R. French and J. Hardisty, (eds.), *Turbulence: Perspectives on Flow and Sediment Transport.* John Wiley, Chichester, 61–92.

Best, J.L, and Ashworth, P.J. 1997: 'Scour in large braided rivers & the recognition of sequence stratigraphic boundaries', *Nature,* 387, 275–277.

Best, J.L. and Reid, I. 1984: 'Separation zone at open channel junctions'. *Journal of Hydraulic Engineering,* 110, 1588–1594.

Best, J.L. and Roy, A.G. 1991: 'Mixing-layer distortion at the confluence of channels of different depth'. *Nature,* 350, 6367, 411–413.

Biron, P.M., De Serres, B., Roy, A.G. and Best, J.L. 1993a: 'Shear layer turbulence at an unequal depth channel confluence'. In N.J. Clifford, J.R. French and J. Hardisty, (eds.), *Turbulence: Perspectives on Flow and Sediment Transport.* John Wiley, Chichester, 197–213.

Biron, P.M., Best, J.L. and Roy, A.G. 1996a: 'Effects of bed discordance on flow dynamics at open channel confluences'. *Journal of Hydraulic Engineering,* 122, 676–682.

Biron, P.M., Roy, A.G., Best, J.L. and Boyer, C.J. 1993b: 'Bed morphology and sedimentology at the confluence of unequal depth channels'. *Geomorphology,* 8, 115–129.

Biron, P.M., Roy, A.G. and Best, J.L. 1996b: 'Turbulent flow structure at concordant and discordant open-channel confluences'. *Experiments in Fluids*, 21, 437–446.

Boyer, C. 1996: Turbulence, transport des sédiments et charge de fond et forme du lit à un confluent de cours d'eau naturels. Thèse de doctorat non-publiée, Université de Montréal.

Bradbrook, K.F., Biron, P.M., Lane, S.N., Richards, K.S. and Roy, A.G. 1998: 'Investigation of controls on secondary circulation and mixing processes in a simple confluence geometry using a three-dimensional numerical model'. *Hydrological Processes*, 12, 1371–1396.

Bradbrook, K.F., Lane, S.N. and Richards, K.S. 2000: 'Numerical simulation of three-dimensional, time-averaged flow structure at river channel confluences'. *Water Resources Research*, 36, 2731–2746.

Bradbrook, K.F., Lane, S.N., Richards, K.S., Biron, P.M. and Roy, A.G. 2001: 'Role of bed discordance at asymmetrical river confluences'. *Journal of Hydraulic Engineering*, 127, 351–368.

Bridge, J.S. 1977: 'Flow, bed topography, grain size and sedimentary structure in open channel bends: a three-dimensional model'. *Earth Surface Processes*, 2, 401–416.

Bridge, J.S. and Jarvis, J. 1982: 'The dynamics of a river bend: a study in flow and sedimentary processes'. *Sedimentology*, 29, 499–541.

Carson, M.A. and Lapointe, M.F. 1983: 'The inherent asymmetry of river meander platform'. *Journal of Geology*, 91, 41–56.

Church, M. 1996: 'Space, time and the mountain – how do we order what we see? In B.L. Rhoads and C.E. Thorn, (eds.), *The Scientific Nature of Geomorphology*, Proceedings of the 27th Binghamton Symposium in Geomorphology, New York, John Wiley, 147–170.

De Serres, B., Roy, A.G., Biron, P.M. and Best, J.L. 1999: 'Three-dimensional structure of flow at a confluence of river channels with discordant beds'. *Geomorphology*, 26, 313–335.

Dietrich, W.E. 1987: 'Mechanics of flow and sediment transport in river bends'. *In* K.S. Richards, (ed.), *River Channels: Environment and Process.* Blackwell, Oxford, 179–227.

Dietrich, W.E. and Smith, J.D. 1983: 'Influence of the point bar on flow through curved channels'. *Water Resources Research*, 19, 1173–1192.

Dietrich, W.E. and Smith, J.D. 1984: 'Bedload transport in a river meander'. *Water Resources Research*, 20, 1355–1380.

Dietrich, W.E., Smith, J.D. and Dunne, T. 1979: 'Flow and sediment transport in a sand bedded meander'. *Journal of Geology*, 87, 305–315.

Drake, T.G., Shreve, R.L., Dietrich, W.E., Whiting, P.J. and Leopold, L.B. 1988: 'Bedload transport of fine gravel observed by motion-picture photography'. *Journal of Fluid Mechanics*, 56, 559–575.

Ferguson, R.I. 1973: 'Channel pattern and sediment type'. *Area*, 5, 38–41.

Ferguson, R.I. 1975: 'Meander irregularity and wavelength estimation'. *Journal of Hydrology*, 26, 315–333.

Gaudet, J.M. and Roy, A.G. 1995: 'Effects of bed morphology on flow mixing length at river confluences'. *Nature*, 373, 138–139.

Giddens, A. 1981: *A Contemporary Critique of Historical Materialism*. Macmillan, London.

Gilbert, G.K. 1877: 'Report on the geology of the Henry Mountains'. U.S. Geological Survey.

Harré, R. 1983: *Personal Being*. Blackwell, Oxford.

Harré, R. 1986: *Varieties of Realism*. Blackwell, Oxford.

Harrison, S. and Dunham, P. 1998: *Transactions of the Institute of British Geographers*, 24,

Hickin, E.J. 1974: 'The development of river meanders in natural river channels'. *American Journal of Science*, 274, 414–442.

Hickin, E.J. and Nanson, G.C. 1975: 'The character of channel migration on the Beatton River, northeast British Columbia, Canada'. *Geological Society of America Bulletin*, 86, 487–494.

Horton, R.E. 1932: 'Drainage basin characteristics'. *Transactions of the American Geophysical Union*, 13, 350–361.

Horton, R.E. 1945: 'Erosional development of streams and their drainage basins: hydrophysical approach to quantitative morphology'. *Geological Society of America Bulletin*, 56, 275–370.

Howard, A.D. 1971: 'Optimal angles of stream junction: geometric stability to capture, and minimum power criteria'. *Water Resources Research*, 7, 863–873.

Kennedy, B.A. 1984: 'On Playfair's law of accordant junctions'. *Earth Surface Processes and Landforms*, 9, 291–307.

Lane, S.N. 2001a: 'The measurement of gravel-bed river morphology'. In P.M. Mosley (ed.), *Gravel-bed Rivers V*, New Zealand Hydrological Society, Wellington, 291–338.

Lane, S.N. 2001b: 'Constructive comments on D. Massey: space-time, "science" and the relationship between physical geography and human geography'. *Transactions of the Institute of British Geographers*, 26, 243–56.

Lane, S.N., Biron, P.M., Bradbrook, K.S., Chandler, J.H., Crowell, M.D., McLelland, S.J., Richards, K.S. and Roy, A.G. 1998: 'Three-dimensional measurement of river channel flow processes using acoustic Doppler velocimetry'. *Earth Surface Processes and Landforms*, 23, 1247–1267.

Lane, S.N., Bradbrook, K.F., Richards, K.S., Biron, P.M. and Roy, A.G. 2000: 'Secondary circulation cells in river channel confluences: measurement artefacts or coherent flow structures?', *Hydrological Processes*, 14, 2047–2071.

Lane, S.N., Chandler, J.H. and Richards, K.S. 1994: 'Developments in monitoring and terrain modelling small-scale river-bed topography'. *Earth Surface Processes and Landforms*, 19, 349–368.

Lane, S.N. and Richards, K.S. 1997: 'Linking river channel form and processes: time, space and causality revisited'. *Earth Surface Processes and Landforms*, 22, 249–260.

Leclair, S. and Roy, A.G. 1997: 'Variabilité de la morphologie et des structures sédimentaires du lit d'un confluent de cours d'eau discordant en période d'étiage'. *Géographie physique et Quaternaire*, 51, 2, 125–139.

Leopold, L.B. and Wolman, M.G. 1957: 'River channel patterns – braided, meandering and straight'. *United States Geological Survey,* Professional Paper 282A.

Massey, D. 1999: 'Space-time, "science" and the relationship between physical geography and human geography'. *Transactions of the Institute of British Geographers*, 24, 261–276.

McLelland, S.J., Ashworth, P.J. and Best, J.L. 1996: 'The origin and downstream development of coherent flow structures at channel junctions'. In P.J. Ashworth, S.J. Bennett, J.L. Best and S.J. McLelland (eds.), *Coherent Flow Structures in Open Channels*. John Wiley, Chichester, 461–489.

Mosley, M.P. 1976: 'An experimental study of channel confluences'. *Journal of Geology*, 84, 535–562.

Paola, C. 1996: 'Incoherent structure: turbulence as a metaphor for stream braiding'. In P.J. Ashworth, S.J. Bennett, J.L. Best and S.J. McLelland (eds.), *Coherent Flow Structures in Open Channels*. John Wiley, Chichester, 705–724.

Playfair, J. 1802: *Illustration of the Huttonian Theory of the Earth*. Cadell and Davies and William Creech, Edinburgh.

Rhoads, B.L. 1996: 'Mean structure of transport-effective flows at an asymmetrical confluence when the main stream is dominant'. In P.J. Ashworth, S.J. Bennett, J.L. Best and S.J. McLelland (eds.), *Coherent Flow Structures in Open Channels*. John Wiley, Chichester, 491–517.

Rhoads, B.L. and Kenworthy, S.T. 1995: 'Flow structure at an asymmetrical stream confluence'. *Geomorphology*, 11, 273–293.

Rhoads, B.L. and Kenworthy, S.T. 1998: 'Time-averaged flow structure in the central region of a stream confluence'. *Earth Surface Processes and Landforms*, 23, 2, 171–191.

Rhoads, B.L. and Sukhodolov, A.N. 2001: 'Field investigation of three-dimensional flow structure at stream confluences: 1. Thermal mixing and time-averaged velocities'. *Water Resources Research*, 37, 9, 2393–2410.

Rhoads, B.L. and Thorn, C.E. 1996: *The Scientific Nature of Geomorphology*. John Wiley, Chichester.

Richards, K.S. 1996: 'Samples and cases: Generalisation and explanation in geomorphology'. In B.L. Rhoads and C.E. Thorn (eds.), *The Scientific Nature of Geomorphology*, Proceedings of the 27th Binghamton Symposium in Geomorphology, New York, John Wiley, Chichester, 171–190.

Richardson, W.R.R. and Thorne, C.R., 1998: 'Secondary currents and channel changes around a braid bar in the Brahmaputra River, Bangladesh'. *ASCE Journal of Hydraulic Engineering*, 124, 325–8.

Rodriguez-Iturbe, I. and Rinaldo, A. 1997: *Fractal River Basins*. Cambridge University Press, New York.

Rosovskii, I.L. 1954: *Concerning the Question of Velocity Distribution in Stream Bends*. SAN URSR (Report of the Academy of Sciences of the Ukraine SSR), 1.

Roy, A.G. 1983a: 'Optimal angular geometry models of river branching'. *Geographical Analysis*, 15, 87–96.

Roy, A.G. 1983b: 'Optimal models of river branching angles'. In M.J. Woldenberg (ed.), *Models in Geomorphology*, The Binghamton Symposia in Geomorphology International Series, no. 14, Allen and Unwin, Boston.

Roy, A.G., Biron, P.M., Buffin-Bélanger, T. and Levasseur, M. 1999: 'Combined visual and quantitative techniques in the study of natural turbulent flows'. *Water Resources Research*, 35, 871–877.

Roy, A.G. and Buffin-Bélanger, T. 2001: 'Advances in the sudy of turbulent flow structures in gravel-bed rivers'. In P.M. Mosley (ed.), *Gravel-bed Rivers V*, New Zealand Hydrological Society, Wellington, 375–404.

Roy, A.G., Buffin-Bélanger, T. and Deland, S. 1996: 'Scales of turbulent coherent flow structures in a gravel bed river'. In P.J. Ashworth, S.J. Bennett, J.L. Best and S.J. McLelland (eds.), *Coherent Flow Structures in Open Channels*. John Wiley, Chichester, 147–164.

Roy, A.G. and De Serres, B. 1989: 'Morphologie du lit et dynamique des confluents de cours d'eau'. *Bulletin de la Société Géographique de Liège*, 25, 113–127.

Roy, A.G. and Woldenberg, M.J. 1986: 'A model for changes in channel form at a river confluence'. *Journal of Geology*, 94, 401–411.

Schumm, S.A. 1960: 'The shape of alluvial channels in relation to sediment type'. *USGS Professional Paper*, 352B, 17–30.

Schumm, S.A. and Lichty, R.W. 1965: 'Time, space and causality in geomorphology'. *American Journal of Science*, 263, 110–119.

Spedding, N. 1997: 'On growth and form in geomorphology'. *Earth Surface Processes and Landforms*, 22, 261–266.

Strahler, A.N. 1952: 'Dynamic basis of geomorphology'. *Geological Society of America Bulletin*, 63, 923–938.

Sukhodolov, A.N. and Rhoads, B.L. 2001: 'Field investigation of three-dimensional flow structure at stream confluences'. *Water Resources Research*, 37, 9, 2411–2424.

Taylor, E.H. 1944: 'Flow characteristics at open channel junctions'. *Transactions of the American Society of Civil Engineers*, 109, Paper 223, 893–912.

Thorne, C.R. and Hey, R.D. 1979: 'Direct measurements of secondary currents at a river inflexion point'. *Nature*, 280, 226–228.

Webber, N.B. and Greated, C.A. 1966: 'An investigation of flow behaviour at the junction of rectangular channels'. *Proceedings and Papers of the Institute of Civil Engineers*, Paper 6901, 321–334.

Westaway, R.M., Lane, S.N. and Hicks, D.M. (in press). 'Remote survey of large-scale braided rivers using digital photogrammetry and image analysis'. Paper forthcoming in *International Journal of Remote Sensing*.

7

Simple at heart? Landscape as a self-organizing complex system

David Favis-Mortlock and Dirk de Boer

> *Why, with the times, do I not glance aside*
> *to new-found methods and to compounds strange?*
> William Shakespeare (c. 1598), Sonnets 76

7.1 The complexity of landscapes

7.1.1 Real landscapes

Here is an exercise for any physical geographer: stroll in the landscape of your choice, look around carefully, and then ask candidly, 'How well can I explain what I see?' OK, it certainly is possible to give a broad-brush explanation. But why is this hollow just *here* and that hummock just *there*? Explanation at this level of detail almost always eludes us.

Albert Einstein suggested that 'The supreme task of the physicist is to arrive at those universal elementary laws from which the cosmos can be built up by pure deduction' (Calaprice, 2000, 256). Geomorphology is a younger science than physics (Davies, 1969), and at this stage of its development no current geomorphologist would seriously suggest that we should be able to deduce the form of a particular landscape from first principles (cf. Harvey, 1969). Nonetheless, it is still instructive (and humbling) to ponder the wide gulf between what we know about the processes that shape the Earth's surface and any landscape's enormous richness and variety of form; its large-scale patterns and its intricate microscale detail; and its sometimes explainable and sometimes apparently idiosyncratic landforms.

In short, landscapes are complex. What this means in practical terms is that we cannot readily predict the effect of a disturbance: for example, how an increase in annual precipitation will affect a river's sediment load and change the morphology of its channel. We also cannot easily deduce how changes in past climate may have affected this river. The reason for this, we suggest here, is that in a landscape one always has to deal with very many components, the interaction of which[1] determines how the landscape functions, evolves and responds to disturbances. This notion of 'complexity' goes beyond

---

[1] As well as those interacting components which operated in the past (but which may no longer operate) and which have left their stamp on the present-day landscape (see 7.4.1.2 below).

that of merely 'complicated' (which describes something that is hard to understand). One recent quantitatively based definition of this kind of complexity is: 'the smaller the parts that must be described to describe the behaviour of the whole, the larger the complexity of the whole system' (Bar-Yam, 1997, 14). Yet Schumm (1991, 48) suggests that it is necessary to look at a landscape at *all* spatial scales in order to understand it. Following this line of thought, landscapes are highly – even infinitely – complex, and a full understanding may be forever beyond our reach.

However, this chapter proposes a more optimistic view. We suggest that while landscapes are indeed very complex, the 'components' of the landscape, and the interactions between these components, may be much simpler. From this perspective, the complexity of landscapes arises out of the enormous number of interactions between these simpler 'components'. We aim to show here how (mainly non-geographical) research from the late 1980s, focusing on self-organization in complex systems, has led to insights which have the potential to lead us to a new view of the landscape, with improved ability to explain the functioning and evolution of landscape systems.[2] We illustrate these insights with results from two models.

7.1.2 Models of landscapes

Mathematical models are the scientific form of analogies (Harvey, 1969). Just as the aim of an analogy is to highlight similarities between the things compared, models of landscapes aim to link the more easily understood system (the model) with the less comprehensible system (the real landscape). But, just as analogies are never perfect, so are models always less than perfect in making this link. Indeed, it could hardly be otherwise, since a model which is as complicated as the thing it represents would be no easier to understand. 'The best material model of a cat is another, or preferably the same, cat' (Rosenblueth and Wiener, 1945).

The nature of the link between model and reality, though, will depend upon the purpose of the model (Haggett and Chorley, 1967). Landscape models are used for a variety of purposes. One class of usage is theoretically oriented: here the objective is to test theories, to organize knowledge and data and to develop new insights and obtain new knowledge of landscapes and how they come to be. More pragmatic usages aim to forecast and predict process rates and landform geometry, both for the present and near future (e.g. for agricultural and conservation-oriented applications) and for long periods of time (e.g. under different scenarios of past and future climate).

Yet, irrespective of the purpose of the model, there has been – and still is – a general trend for landscape models to become more and more complicated. This trend can be traced back to the 'quantitative revolution' in Geography in the 1960s (Harvey, 1969),

---

2 A system is a set of objects and their interactions; the system must 'make sense' as a unit. In geomorphology, a landscape or a part of a landscape may be viewed as a system (Chorley, 1962).

when models first found widespread usage in almost all areas of the subject.[3] The trend shows no signs of abating. To illustrate this, Table 7.1 uses the number of lines of programming code as an indicator (cf. the notion of 'algorithmic complexity', Chaitin, 1999) of how field-scale computer models for soil erosion by water have become more complicated.

Table 7.1 *An estimate of the complexity of some field-scale erosion models using the number of lines of programming source code. The date given is of a 'representative' publication and not the release analysed here. Project Line Counter 1.11 from* http://www.wndtabs.com *was used for analysis of models marked *. From Favis-Mortlock et al., 2001.*

Model	Version	Date	Number of source code lines	Programming language(s)	Comment
Guelph USLE	–	1978	550	Basic	An implementation of the Universal Soil Loss Equation (Wischmeier and Smith, 1978) programmed by D.J. Cook of the University of Guelph, September 1985. CALSL module only considered here
EPIC	3090	1983	10689	Fortran *	USLE-based, but continuous simulation, and includes crop model, weather generator. Williams et al. (1983)
GAMES	4.00	1986	1744	Fortran *	USLE-based. Dickinson et al. (1986)
GLEAMS	2.03	1987	18106	Fortran, C	Continuous simulation. Leonard et al. (1987), based largely upon the earlier CREAMS model (Knisel, 1980)
WEPP	99.5	1989	59031	Fortran	*Continuous simulation and event, combines hillslope and watershed versions, includes crop model. Nearing et al. (1989)
EUROSEM	3.12	1998	5943	Fortran	*Single event only. Morgan et al. (1998)

3 Although models had begun to be used in geology a good deal before this, as evidenced in this pithy quote: 'Somehow I hate to see budding geologists feeding machines numbers. They should be out in the field learning to get the right numbers to put into the infernal machines' (Link, 1954).

This trend results from a natural desire for 'better' models. Despite the wide variation between individual landscape models, there are two general criteria which have come to dominate in assessing a geomorphological model's 'success' or otherwise, particularly in practical applications where the main aim is a successful replication of measured values (often with the subsequent aim of estimating unmeasurable values, e.g. under some hypothetical conditions of climate or land use).

The first criterion is theoretical: the model is usually judged more highly if it is physically based, i.e. if it is founded on the laws of conservation of energy, momentum and mass, which have their origin within physics; and if it has its parameters and variables defined through equations that are at least partly based on the physics of the problem, such as Darcy's law and the Richards equation (Kirkby et al., 1992). A physical basis is considered desirable because the laws of physics are assumed to be universally applicable: thus the more a model is rooted in these laws (and, conversely, the less it depends upon empirically derived relationships), the more widely applicable – i.e. less location-specific – it is assumed to be.

The second criterion is pragmatic: how well the model does in a comparison of model results against measured values. For example, a time series of simulated discharges might be compared with an observed time series, or simulated soil loss compared with measured values (Favis-Mortlock et al., 1996 and Favis-Mortlock, 1998a). Differences between computed and observed time series are normally attributed to the failure of the model to reflect some aspect of the real world. As an illustration of this, in an evaluation of catchment-scale soil erosion models (Jetten et al., 1999), the models were largely able to correctly predict sediment delivery at the catchment outlet, but were much less successful at identifying the erosional 'hot spots' within the catchment which supplied this sediment. The failure was in part attributed to deficiencies in modelling within-catchment flow paths (see also Takken et al., 1999 and Steegen et al., 2000).

The remedy for such shortcomings in models is almost always perceived to be the addition of detail – preferably physics-based – to the model.[4] Thus the empirical Universal Soil Loss Equation (USLE: Wischmeier and Smith, 1978) did not distinguish between rainsplash-dominated inter-rill soil loss and flow-dominated rill erosion. These processes are separately modelled however, in a more physically based way, in two subsequent models: the Water Erosion Prediction Project (WEPP: Nearing et al., 1989) model and the European Soil Erosion Model (EUROSEM: Morgan et al., 1998). Overall, this strategy inevitably leads to an explosion in model complexity and data requirements.

---

4 Physicists distinguish between 'perturbative' and 'non-perturbative' approaches. The idea is that one essentially starts with a guess regarding process description, and gradually refines that guess by adding in more information. If the original guess was reasonable, then this process should gradually refine the original guessed description so that it more and more closely approximates reality. But if adding in more information has a major effect on the guessed description, a rethink is indicated (Greene, 1999). This formalism might usefully be applied to geographical models.

An associated problem stems from our incomplete knowledge of the physics of landscape-forming processes. Current understanding of several of these processes is still relatively weak (e.g. of the details of soil surface crusting), and thus these processes can only be described in current models in a more-or-less empirical way. As a result, some model inputs essentially fulfil the function of curve-fitting parameters[5] that are adjusted to provide a match between the observed and computed time series rather than measured independently. Under such conditions, model parameterization becomes more and more a curve-fitting exercise (Kirchner et al., 1996), so that (for example) a recent evaluation of field-scale erosion models (Favis-Mortlock, 1998a) found calibration to be essential for almost all models involved, despite the avowed physical basis of the models. This development has been much to the dismay of authors such as Klemes, who wrote: 'For a good mathematical model it is not enough to work well. It must work well for the right reasons. It must reflect, even if only in a simplified form, the essential features of the physical prototype' (Klemes, 1986, 178S).

Additionally, if interactions between processes are inadequately represented within the model, results from a more complex model may not necessarily improve upon those of a simpler model (Mark Nearing, personal communication, 1992 and Beven, 1996). Adding more model parameters increases the number of degrees of freedom for the model. Jakeman and Hornberger (1993), for example, found that commonly used rainfall – runoff data contains only enough information to constrain a simple hydrological model with a maximum of four free parameters. Adding extra free parameters to a model means that changes in the value of one input parameter may be compensated by changes in the value of another, so that unrealistic values for individual input parameters may still produce realistic results (in the sense of a close match between the observed and computed time series). Thus the model may be right for the wrong reasons (Favis-Mortlock et al., 2001). An end-point to this is Beven's 'model equifinality', whereby entirely different sets of input parameters can still produce similar model outputs (Beven, 1989; the concept of equifinality[6] is originally due to Von Bertalanffy, 1968).

Not all landscape modelling, though, aims to compare computed and measured values. In more 'blue-sky' situations, the aim of model usage is to explain, to provide new insights and to generate new knowledge of the system represented in the model. The objective of this kind of modelling is not so much to replicate some feature of the prototype, but rather to generate new ideas about how the prototype works, to

---

5 Cohen and Medley (2000, 32) point out that in classical biological experimentation, a clear distinction is made between parameters and variables. Variables are those properties of the experimental set-up which are within the hypothesis being tested, and whose values will be varied in the course of the experiment. Parameters are those properties of the experimental set-up which are not directly implicated in the hypothesis, but which may still affect results. No such clear distinction is made in geographical modelling.
6 Equifinality is the notion that an infinite number of different sets of assumptions can lead to the same consequences. Similar-appearing landforms resulting from the operation of differing processes are sometimes termed 'homologous' or 'convergent', e.g. Pitty (1971, 12).

attempt[7] to capture the causality of real-world systems. However, if such models are to be humanly comprehensible, then an important consideration is the extent to which the model is a simplification of reality. If the model is too complex, then crucial similarities between the model and the prototype cannot be recognized and emphasized (Flake, 1998). Again, in Albert Einstein's words,[8] 'Things should be as simple as possible, but not any simpler'. That is, the model should be a simplified version of the prototype, but even though it is simplified, it must retain the essential features of the prototype. This view is expressed clearly by Gleick (1987, 278):

> The choice is always the same. You can make your model more complex and more faithful to reality, or you can make it simpler and easier to handle. Only the most naïve scientist believes that the perfect model is the one that perfectly represents reality. Such a model would have the same drawbacks as a map as large and detailed as the city it represents, a map depicting every park, every street, every tree, every pothole, every inhabitant, and every map. Were such a map possible, its specificity would defeat its purpose: to generalize and abstract. Mapmakers highlight such features as their clients choose. Whatever their purpose, maps and models must simplify as much as they mimic the world.

Thus, one cornerstone of the art of modelling lies in knowing or sensing which aspects of the prototype must be retained in the model, and which ones may be omitted. This need for understandable simplicity is clearly impeded by the trend to make landscape models more complicated.

7.2 Non-linear systems: deterministic chaos and self-organization
7.2.1 Deterministic chaos

So is there any way out of this decidedly non-virtuous circle, whereby a 'better' model of the landscape inexorably has to describe more processes, and in so doing requires more data and becomes less comprehensible and thus less informative? More fundamentally, we might ask: does real-world complexity of form always imply an equivalent complexity of the underlying generative processes? A clue that this might not always be so appeared in the 1960s, with the recognition of the phenomenon of deterministic chaos by a number of workers, including the atmospheric physicist Edward Lorentz (Ruelle, 2001). While its mathematical roots date back considerably further (in particular to the work of mathematician Henri Poincaré around the beginning of the 20th

---

7 However, we can never know the causality of the real-world systems we observe, as David Hume (1711–1776) observed. All we can do is to measure how well our subjectively generated models correlate with the behaviour of these systems.

8 This may not actually be due to Einstein: see p314 of Calaprice (2000). At any rate, this notion has become known as 'Einstein's Razor', cf. 'Occam's Razor' as discussed in 7.4.1.3 below)

century: Jones, 1991), deterministic chaos only entered the scientific mainstream in the 1970s,[9] for example, with the work of biologist Robert May on chaotic population dynamics (May, 1976).

One aspect of deterministic chaos that is relevant to landscape modellers is summed up in the colourful question:[10] 'Does the flap of a butterfly's wings in Brazil set off a tornado in Texas?' In other words, can a large effect have a tiny cause? Such non-linear 'extreme sensitivity to initial conditions' is indeed a hallmark of chaotic systems. Thus small uncertainties in measurement or specification of initial conditions become exponentially larger, and the eventual state of the system cannot be predicted. So, for example, the chaotic component of the earth's atmosphere means that weather forecasts rapidly diminish in reliability as one moves more than a few days into the future. Notice, though, that whereas weather (i.e. the particular set of meteorological conditions on a specific day at a specific location) *cannot* be predicted, climate (the range of meteorological conditions during a particular short period, at that location) *can* be predicted. This notion is at the heart of recently devised 'ensemble forecasting' techniques which are carried out using atmospheric models (Washington, 2000).

Another relevant aspect is the close association between deterministic chaos and fractal (i.e. scale-independent) pattern formation. Following seminal work by Benoit Mandelbrot (1975; see also Andrle, 1996), fractal patterns were acknowledged to be present in a wide variety of geographical situations (but see Evans and McClean, 1995). The linkage between fractals and systems exhibiting deterministic chaos suggested some deeper connection (cf. Cohen and Stewart, 1994).

In the present context, the most interesting insight from chaotic systems is that they do not have to be complicated to produce complex results. Lorentz's atmospheric model comprised only three non-linear equations, and May's population dynamics models were even simpler. In all such models, the results at the end of one iteration (in the case of a time-series model, at the end of one 'timestep') are fed back into the model and used to calculate results for the next iteration. This procedure produces a feedback loop. For particular values of the model's parameters, the model's output will eventually[11] settle down to a static equilibrium value; for other values, the model may eventually settle down to cycle forever between a finite number of end-point output values; but for others, the model will switch unpredictably between output values in an apparently random way.

Thus in such chaotic systems, complex patterns can be the results of simple underlying relationships. This discovery has its sobering aspect, representing as it does the final death

---

9 Interestingly, there is some evidence for two waves of interest in chaos theory, each with a distinct disciplinary bias. See http://real.geog.ucsb.edu/pub/chaos/
10 Originally the title of a 1970s lecture by Lorentz. There are now many variants.
11 Note the word 'eventually'. The repetitive calculations which are often necessary are ideally suited to a computer, but not to a human. This is one reason why deterministic chaos had to wait for the widespread usage of computers for its discovery.

rattle of the notion of a predictable, 'clockwork' universe, even at the macroscale.[12] But there is also a strongly positive implication: complexity does not have to be the result of complexity!

Yet however intriguing and suggestive was this early work on deterministic chaos, it was not immediately 'useful' for landscape modellers. For one thing, while the output from chaotic functions is complex, it includes little in the way of immediately recognizable structure: at first glance it more resembles random noise. It is therefore qualitatively very different from the highly structured patterns which we observe in landscapes. For another, when analysing real-world measurements which plausibly possess a chaotic component, it has proved to be very difficult to go from the data to a description of the underlying dynamics (Wilcox et al., 1991), even though simple models may be devised which produce simulated data closely resembling the measured values. But since there may be many such models, what certainty can we have that any one of them has captured the workings of the real landscape? (Again, cf. Beven's 'model equifinality', Beven, 1989.) So, do studies of deterministic chaos represent something of a dead end for physical geographers?

7.2.2 Self-organizing systems

Work on deterministic chaos thus suggested that complex patterns, of some kinds at least, do not always require an underlying generative complexity. The hint became both broader and stronger as a result of pioneering research on self-organization in complex systems. This was carried out from the late 1980s. A major centre for this work was the Santa Fe Institute (Waldrop, 1994). Overviews of this diverse body of early research on self-organization, emergence and complexity are presented by Coveney and Highfield, (1995), Kauffman (1995) and Holland (1998), among others.

Such studies demonstrate that responses which are both complex and highly structured may result from relatively simple interactions between the components of the non-linear system under consideration. Interactions between such components are governed by 'local' rules, but the whole-system ('global') response is to manifest some higher-level 'emergent' organization, following the formation of ordered structures within the system. The system thus moves from a more uniform ('symmetrical') state to a less uniform – but more structured – state: this is so-called 'symmetry-breaking'. Some attributes of the phenomenon of self-organization are summarised in Figure 7.1. Note the definition of emergence which is given: 'emergent responses cannot be simply inferred from the behaviour of the system's components'.

---

12 As Berry (1988) puts it: 'But what we've realised now is that unpredictability is very common, it's not just some special case. It's very common for dynamical systems to exhibit extreme unpredictability, in the sense that you can have perfectly definite equations, but the solutions can be unpredictable to a degree that makes it quite unreasonable to use the formal causality built into the equations as the basis for any intelligent philosophy of prediction'.

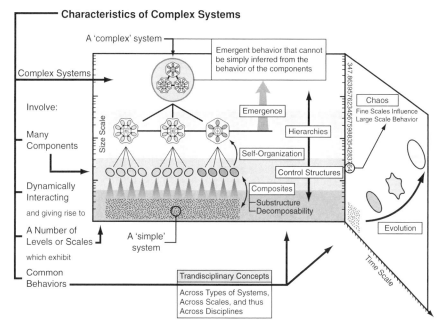

Characteristics of Complex Systems

A 'complex' system

Emergent behavior that cannot be simply inferred from the behavior of the components

Complex Systems

Involve:

Many Components

Dynamically Interacting

and giving rise to

A Number of Levels or Scales

which exhibit

Common Behaviors

Emergence

Hierarchies

Self-Organization

Control Structures

Composites

Substructure
Decomposability

A 'simple' system

Trandisciplinary Concepts

Across Types of Systems, Across Scales, and thus Across Disciplines

Size Scale

347.863957623456759883542836

Chaos
Fine Scales Influence Large Scale Behavior

Evolution

Time Scale

Figure 7.1 *Schematic representation of some attributes of self-organizing complex systems. Graphic by Marshall Clemens (http://idiagram.com) from the 'Visualizing Complex Systems Science' project at http://necsi.org/projects/mclemens/viscss.html. Used by permission.*

7.2.2.1 Characteristics of self-organizing systems

Three of the most important concepts that characterize self-organizing systems are:

- feedback
- complexity and scale
- emergence.

All are interlinked.

Feedback

Both positive and negative feedback can be illustrated by considering a device in a humanly engineered system, the steam engine. Steam engines commonly possess a 'governor'. This is an arrangement whereby a valve is controlled by weighted arms which revolve, the revolution being driven by the steam engine. As the arms rotate faster, increased centrifugal force makes them close the valve somewhat and so slow the engine. As the engine slows, the arms of the governor also rotate more slowly and so open the valve a little; the engine thus speeds up. It is the dynamic balance between the opening of the valve ('positive feedback') and the closing of the valve ('negative feedback') which enables the steam engine to maintain a constant speed. This interaction between positive and negative feedback permits the steam engine to process data about its own state.

In general, negative feedback in a system occurs when the system functions in such a way that the effects of a disturbance are counteracted over time, bringing the system back to its pre-disturbance state. In landscapes,[13] an example of negative feedback occurs when deposition of coarse sediment takes place in a channel section. The resulting increase in gradient at the downstream end of the deposit is reflected in an increase in flow velocity and bed shear stress, ultimately resulting in increased erosion of the channel bed and removal of the coarse sediment deposit, and a return to the earlier conditions.

Positive feedback takes place when a disturbance continues to force the system away from its earlier state. Whereas negative feedback counterbalances change and drives the system back to the pre-disturbance conditions, positive feedback reinforces change and may lead to an entirely new state of the system. An example of positive feedback can occur when soil erosion exposes an underlying soil horizon with a low permeability, which reduces the infiltration rate. Consequently, the rate of overland flow production increases and, in some cases, erosion increases as a result.[14] Chorley et al. (1984, 10) remark that positive feedback in a landscape is not limited to short time spans and small areas, and discuss the role of positive feedback in landscape evolution at large temporal and spatial scales over which landscapes undergo progressive change.

In self-organizing systems, feedback plays a crucial role in the formation of spatial and temporal patterns. An example of the role of positive feedback in landscape evolution occurs when erosion locally lowers the surface, resulting in an increased concentration of water and still more erosion, until at some point a valley is formed which ultimately becomes part of a larger channel network. Viewed in this manner, the formation of a drainage network on a continental surface is the direct result of positive feedback that enhances initially minor differences in surface elevation. At the same time, however, the occurrence of negative feedback prevents the valleys from becoming overly deep, by the deposition of sediment in any low spots in the drainage network. Thus a balance is maintained: while 'successful' channels (see 7.3.2.1 below) are deepened, no channel can become too deep too quickly. It is this non-linear recursive interplay[15] which leads to the emergence of channel networks (see 7.3.2 below).

---

13 According to Pitty (1971, 71), the notion of feedback was introduced to British geomorphology in the 1960s via Phillipe Pinchemel's work on dry valleys in chalk.
14 This increase in runoff will not always result in increased erosion since in some cases the exposed B horizon will have greater shear strength and so be better able to resist detachment by the extra flow.
15 Referred to as 'complicity' in Cohen and Stewart, 1994. As Jack Cohen put it in an email to complex-sciences@necsi.org on 10th December 2000: '... a more exquisite problem for complex thinking is the problem of recursion. The system doesn't simply respond to the environment, the environment also responds to the system. The pattern of a river bed is the result of genuine interaction of this kind, and so is nearly every ongoing process in physics, biology or management'. Another term for this is 'autocatalysis'.

Complexity

In order to exhibit self-organization, a system must be complex i.e. must possess sufficient scope[16] for component-level interactions (which can be characterized as positive and negative feedback) to give rise to system-wide, emergent responses. However, while all self-organizing systems are complex, not all complex systems are self-organizing (Çambel, 1993, 20). At present though, there is no single, precise definition of complexity (Bar-Yam, 1997 and Chaitin, 1999). Gallagher and Appenzeller (1999) loosely define a complex system as a system with properties that cannot be described fully in terms of the properties of its parts. Most authors do not provide a single definition of complexity, but instead describe various characteristics of complexity.

The notion of complexity is closely tied to our conceptualization of scale. Bar-Yam (1997, 258) points out that, 'The physics of Newton and the related concepts of calculus, which have dominated scientific thinking for three hundred years, are based upon the understanding that at smaller and smaller scales – both in space and in time – physical systems become simple, smooth and without detail'. From this viewpoint, the assumption is that even the most complex of systems, when viewed at a 'component scale', somehow[17] becomes simpler, and thus more mathematically tractable. This assumption is at the heart of much of present-day mathematical modelling. The assumption is true for some systems,[18] but for those systems which self-organize, to ignore the complex within-system interactions which give rise to that self-organization is to throw the baby out with the bathwater.

Emergence

The roots of the notion of emergence go back at least to *c.* 330 BCE with Aristotle's description of synergy: 'The whole is more than the sum of its parts'. An emergent response is synergistic, but 'more so' (in a qualitative sense).

The continual flows of energy and matter through a dissipative system (see 7.2.2.3 below) maintain it in a state far from equilibrium (Ruelle, 1993 and Çambel, 1993). Ordered structures 'emerge' as a result of interactions between the system's sub-components, such interactions being driven by the flow of matter and energy which characterizes such systems. As these structures grow more common within the system, the system as a whole 'self-organizes'. This transition (see 'symmetry-breaking' above) often occurs quite rapidly and abruptly, in the manner of a phase change e.g. from water to ice. It is crucial to note that this increase in systemic organization is entirely a result of

---

16 In other words, a large number of independently varying degrees of freedom.
17 e.g. because of the averaging effect of the Law of Large Numbers.
18 Such as an idealized gas: see Bar-Yam (1997). It is also true (but in a different way) for almost all systems, even quite complicated ones, which have been constructed by some external agency e.g. by a human. A first question to ask when attempting to identify self-organization is: 'Is this system more than the sum of its parts?'

internal interactions, rather than resulting from some externally imposed controlling factor (although a flow of energy and matter through the system is essential).

Bar-Yam (1997, 10) points out that even in the scientific community there is confusion regarding the nature of emergence. One 'fingerprint' of emergence is that the emergent properties of a system cannot easily be derived from the properties of the system components or sub-systems (see Figure 7.1). Additionally, Bar-Yam (1997, 10–12) distinguishes between local and global emergence. In local emergence, the emergent response of the system is relatively resistant to local perturbations: for example, a standing wave on the lip of a waterfall will remain, despite innumerable small variations in flow conditions in its vicinity. In global emergence, the emergent response of the system is more susceptible to local perturbations: thus any living thing will be adversely affected by disturbance of even small parts of its body. Local and global emergence clearly form end-points in a continuum.

7.2.2.2 Self-organized criticality

Studies by Bak and co-workers (Bak et al., 1988 and Bak, 1996) on sand-pile models have provided a number of insights into other generic aspects of complex systems. While still controversial, this work on so-called 'self-organized criticality' (SOC) suggests that the presence of power-law frequency-magnitude relationships, 1/f properties of time-series data, and spatial fractality form a kind of fingerprint for SOC, and hence for self-organization in the more general sense (Bak, 1996 and Buchanan, 2000). This view suggests that self-organization can be manifested not only in emergent pattern formation, but also in terms of particular configurations of the internal dynamics of systems.

7.2.2.3 Intuition, entropy and self-organizing systems

Self-organizing systems have been proposed or identified all around us:

- in purely physical systems[19] e.g. crackling noise (Sethna et al., 2001), chemical reactions, Tam, 1997) and sand-dune dynamics (Hansen et al., 2001)

- in biological systems e.g. population dynamics (Solé et al., 1999), embryonic pattern formation (Goodwin, 1997), ecology (Laland et al. 1999), evolutionary dynamics (Lewin, 1997) and Gaia Theory (Lenton, 1998)

- in human systems e.g. urban structure (Portugali et al., 1997), social networks (Watts, 1999) and the discipline of Geography (Clifford, 2001).

Yet since our common experience is that bricks do not spontaneously organize themselves into houses, or a child's bedroom spontaneously tidy itself, at first acquaintance the notion of self-organization is distinctly counter-intuitive (Bar-Yam, 1997, 623).

---

19 Samuel (2002) even suggests that the fundamental building blocks of matter could be emergent: 'It may be that what we call reality is a spontaneous phenomenon, emerging like a wave out of some forever unknowable cosmic medium'.

We suggest that this is, to some extent, a matter of preconceptions. It is a matter of everyday observation that the hot cappuccino always cools to muddy uniformity. It is perhaps this kind of much-repeated experience which colours our expectations of 'the way things are'. But equally, we see living things – plants and animals – come into being, and maintain their organization for some time (i.e. as long as they are alive). Accustomed to this since infancy, we unquestioningly intuit that these 'kinds' of system are different in some fundamental way (although we may seldom ask ourselves what this difference is). So why do these two kinds of systems behave so differently?

Distinctions between living and non-living systems were much discussed at the end of the 19th century, when the laws of thermodynamics were being formulated by physicists such as Ludwig Boltzmann. These laws express some of the most profound scientific truths yet known, in particular, the universal inexorability of dissolution and decay. To develop an understanding of self-organization, it is essential to comprehend the implications of the second law of thermodynamics. This states that entropy, or disorder, in a closed system can never decrease: it can be colloquially expressed as 'There is no such thing as a free lunch'. Thus in a 'closed' finite universe, the end result is a kind of cold-cappuccino uniformity, often described in studies of chemical equilibrium (Chorley, 1962). However, pioneering work by Ilya Prigogine and colleagues at the Free University of Brussels in the 1960s (Nicolis and Prigogine, 1989; Ruelle, 1993 and Klimontovich, 2001) focused on systems which are 'open' from a thermodynamic perspective. In a thermodynamically open system (also called a 'dissipative system': Çambel, 1993, 56), there is a continual flow of matter and energy through the system. This continuous flow of matter and energy permits the system to maintain itself in a state far from thermodynamic equilibrium (Huggett, 1985) – at least while the flow continues. This is in contrast to the thermodynamically closed (or 'conservative') system, such as the cup of cappuccino together with its immediate surroundings: here there is no such flow of matter and energy into the system, only a movement of energy between the coffee and its surroundings. Thus the coffee and its environment gradually equilibrate, with the coffee cooling and mixing, and the surrounding air warming slightly. Living systems, though, maintain their structure and do not equilibrate, as long as food and oxygen etc. are available. So in a sense, the second law of thermodynamics defines the current against which all living things successfully swim (Wright, 2000, 159), while they are alive. 'Ever compare a five-course meal with the resulting excrement? Something has been lost … But something has been gained too … the secret of staying alive … is to hang on to the order and expel the disorder.' (Wright, 2000, 244).

All self-organizing systems, living or non-living, are alike in being thermodynamically open. This is not a new insight for geographers: for example, Leopold et al. (1964, 267) note that dynamic equilibrium of fluvial systems refers:

> *to an open system in which there is a continuous inflow of materials, but within which the form or character of the system remains unchanged. A biological cell is such a system. The river channel at a particular location over a period of time similarly receives an inflow of sediment and water, which is discharged downstream, while the channel itself remains essentially unchanged.*

Thus each non-living dissipative system must, as any living system must maintain a flow of matter and energy through itself in order to retain its integrity. Without this, it becomes a thermodynamically closed system, with bland equilibration as its only future. While our intuition correctly distinguishes between closed and dissipative systems when the dissipative system is living, it does not reliably distinguish between closed and dissipative systems when the dissipative system is non-living. Non-living dissipative systems that organize themselves are therefore a surprise to us.[20]

7.2.2.4 Modelling self-organization: CA models

If a real system manifests self-organization and emergence, then one way to study these is by means of a model which also manifests these phenomena. A number of approaches have been developed for modelling self-organizing complex systems. For spatial systems such as landscapes, one of the most useful is the 'cellular automaton' (CA) approach.

CA models discretize continuous space[21] into a series of cells, which are usually part of a regular square or rectangular grid. This spatial discretization[22] is necessary if the model is to be computationally tractable. The rules and relationships which comprise the model are then applied at the scale of individual cells. These rules and relationships may be viewed as positive and negative feedback, and thus each cell may be seen as an 'automaton' with its behaviour controlled by the positive and negative feedback which comprise the model. Interactions are usually (but not always) confined to those between adjacent or nearby cells: thus interactions are all 'local'. If the CA model then self-organizes and gives rise to larger-scale responses, these will manifest as patterns on the cellular grid (Mahnke, 1999 and Wooton, 2001). Watts (1999, 181 *et seq.*) gives a good overview of the history and development of CA models. They are now widely used in the study of self-organization in a wide range of fields (Bar-Yam, 1997). A simple CA model is the 'Game of Life' (cf. Wolfram, 1982), in which a limited number of individual cells start out 'alive'. Each timestep, these cells either 'live' (i.e. persist) or 'die' (i.e. are

---

20 Yet from still another point of view, they certainly shouldn't be. For if non-living systems do not organize themselves, how would they get organized in the first place? Unless we invoke an organizer (i.e. teleology), then from this perspective 'if self-organization did not exist, it would be necessary to invent it'.
21 Just as almost all models which operate in a true time dimension discretize continuous time into distinct 'time steps'.
22 Discretization appears to be a fundamental operation for humans attempting to make sense of the world: see Bohm (1980) chapter one.

removed), depending on the number of surrounding cells. As simple as this may appear, when applied iteratively some very complex patterns can result. Overall, the grid may evolve to a steady state, to some iterating pattern or to an apparently disordered configuration (cf. the chaotic systems described in 7.2.1 above).

Several CA models are freely available on the internet.[23] Note that whereas the rules in the 'Game of Life' CA model are deterministic, this is not necessarily the case, particularly in more complicated CA models, such as the two landscape models described in this chapter. Similarly, the original configuration of the CA grid (i.e. at the beginning of the simulation) may be arrived at by deterministic or stochastic means[24]. A variant of the simple CA model, which is perhaps more suited to representing continuum systems such as fluvial flow' is the 'lattice-gas model' (Wolfram, 1986; Garcia-Sanchez et al., 1996; Pilotti and Menduni, 1997; see also http://poseidon.ulb.ac.be/lga_en.html).

Bar-Yam (1997, 490) suggests that 'The idea of a cellular automaton is to think about simulating the space rather than the objects that are in it', and (1997, 139) that 'Cellular automata are an alternative to differential equations for the modelling of physical systems'. While the same real-world system may be described either by a CA model or by a more conventional approach (e.g. a fluid dynamics model for a fluvial system), it appears to be the case that the CA approach usually brings out different characteristics of the system when compared with the more conventional representation.[25]

Finally, a note on terminology. There is a tendency by some geomorphological authors (Smith, 1991 and Coulthard et al., 2002) to discuss CA models rather than self-organization and emergence. However, CA models, when used to represent some real-world system, are just tools for reproducing the self-organization and emergence which is assumed to also manifest in the real-world system. Thus, an emphasis on the CA model itself is misleading, since the model is just a means to explore the self-organization and emergence of the original system. It is self-organization and emergence that are the deeper concepts and which better deserve our attention.

7.2.2.5 Modelling self-organization: the problem of context and boundaries

'If everything in the universe depends on everything else in a fundamental way, it may be impossible to get close to a full solution by investigating parts of the problem in isolation' (Hawking, 1988). As Stephen Hawking acknowledges, it is increasingly recognized that reductionism, i.e. breaking apart many systems – including those of interest to geographers – into smaller, more tractable units, poses a risk for full understanding. The reductionist's concentration on the components of a self-organizing system, away from the context in

---

23 e.g. http://psoup.math.wisc.edu/Life32.html, http://sero.org/sero/homepages/, http://www.automaton.ch/, http://www.exploratorium.edu/complexity/CompLexicon/automaton.html.
24 See http://www.santafe.edu/~hag/class/class.html for a hierarchical classification of cellular automata.
25 This is related to the concepts of 'trivial duality' and 'non-trivial duality' in physics (Greene, 1999, 297 *et seq.*). If each of two different descriptions of the same system tell us something that the other does not, this is 'non-trivial duality'.

which such components interact and give rise to emergent self-organization, will miss vital points about the way that the system works. Even with a more holistic focus, the imposition of an artificial boundary between the system's components will constrain the interactions between components in the region of the boundary, with potentially strong effects on emergent responses of the system.[26] This is notably the case for CA models of self-organizing systems, where the 'problem of boundary conditions' may be severe. We must set boundaries, but doing so conceptually breaks some of the model's connections to the 'outside world', and so can result in a distorted model (Bar-Yam, 1997, 8; see also 7.3.2 below). So in the final analysis, we must accept that even the best possible model of a self-organizing system remains incomplete.

7.2.3 Self-organization in landscapes

A landscape can be viewed as an assemblage of systems that each can be subdivided into smaller sub-systems and which, in turn, are part of larger systems (De Boer, 1992). This hierarchy of systems is thermodynamically open, since clear fluxes of energy and matter (predominantly water and sediment) move through these systems at all levels. The relief-forming processes of tectonics and volcanism provide an increase of potential energy within the landscape; in drainage basins, precipitation provides an input of water and of energy; and inputs of solar energy and water drive the processes of weathering and of vegetation growth. These inputs provide the energy that is available at the surface of the earth[27] to move sediment and shape the landforms.

So a landscape fulfils the conditions for a dissipative system (see 7.2.2.3 above). It is also complex. But is it a self-organizing system?

7.2.3.1 Field and laboratory studies of self-organization in landscapes
Early observations
Even decades-old descriptions of landscape-forming processes acknowledge that major consequences could follow from minor causes, as in the following (questionable?) story of the initiation of erosion by Burges (1938):

> In [the Piedmont region of the eastern US] occurs the Providence Gully, known to all students of soil erosion for its immensity. It is probably the deepest man-made gully on [sic] the Western Hemisphere. This gully was started about sixty years ago by the drip from a barn roof. Since then it has swallowed the barn which gave it birth, a school house, a tenant house, and a grave yard with fifty graves. From its present edge it drops downward 200 feet into the bowels of the earth, and in its making has destroyed 50 thousand acres of splendid farmland.

---

26 'Biologists cannot adopt a reductionist approach when working with a living organism, or it dies' (Jack Cohen, personal communication, 2000).
27 Of course, most of the energy in rainfall and runoff is lost as heat to the atmosphere.

The first geomorphological studies

Less dramatic, but more scientific, notions of self-organization in landscapes began to appear in studies from the 1980s (Craig, 1982). During the 1990s, geomorphologists increasingly began to point to specific examples of probable self-organization in the landscape. In some cases these are strikingly expressed in spatial patterns at a variety of scales: from centimetres (e.g. ripples) to thousands of kilometres (e.g. drainage networks covering continents). Hallet (1990) illustrates such spatial patterns in landscapes with many, diverse examples. Other more recent publications have also focused on the role of self-organization in producing patterns in landscapes, (Ball, 1999, Werner, 1999 and Richards et al., 2000), while other such work is still to be published (Higgitt and Rosser, 2002, Turkington and Phillips, 2002).

Self-organized criticality

Other geomorphological studies have focused not so much on spatial patterning as on the power law and 1/f signatures of SOC (see 2.2.2 above). For example, Boardman and Favis-Mortlock (1999) speculated that power-law frequency-magnitude relationships in a wide range of soil loss and sediment yield data might indicate the workings of SOC in the erosional system. This has also been a theme in the work of Phillips (1995; 1996; 1997 and 1999) and Dikau (1999).

7.2.3.2 Self-organizing systems approaches to landscape modelling

Just as field and laboratory workers have been quick to investigate the implications of self-organization for landscape studies, landscape modellers have also eagerly embraced these new concepts, from the early 1990s, with a number of studies using CA models.

First geomorphological modelling studies

Chase (1992) used a cellular model to investigate the evolution of fluvially eroded landscapes over long periods of time and at large spatial scales. In this model, rainfall occurs on the simulated landscape in single increments called precipitons. After being dropped at a random position, a precipiton runs downslope and starts eroding and depositing sediment, depending on the conditions in the cell. Through time, as numerous precipitons modify the landscape, the simple rules gives rise to a complex, fluvially sculpted landscape.

Werner and Hallet (1993) also used the CA approach to investigate the formation of sorted stone stripes by needle ice, and found that the formation of stone stripes and similar features, such as stone polygons and nets, reflects self-organization resulting from local feedback between stone concentration and needle-ice growth rather than from an externally applied, large-scale template. Werner and Fink (1993) similarly simulated the formation of beach cusps as self-organized features: the CA model here was based on the interaction of water flow, sediment transport and morphological change. Werner

(1995) again used the approach to study the formation of aeolian dunes, and Werner (1999) reviewed earlier work and pointed out research directions.

The CA model of Murray and Paola (1994; 1996 and 1997) was developed to investigate the formation of braided channel patterns. Their model replicated the typical dynamics of braided rivers, with lateral channel migration, bar erosion and formation, and channel splitting, using local transport rules describing sediment transport between neighbouring cells. (See also Thomas and Nicholas, 2002.)

Simulation of rill networks formed by soil erosion by water was the goal of the plot-scale RillGrow 1 model constructed by Favis-Mortlock (1996 and 1998b). The model successfully reproduced realistic-appearing rill networks, and was also able to reproduce other 'global' responses such as the relationship between total soil loss and slope gradient.

In addition to focusing on the pattern-forming aspects of self-organization, there has also been a flurry of interest in modelling the more statistical aspects of self-organization such as SOC. Hergarten and Neugebauer (1998 and 1999) investigated the magnitude-frequency distribution of landslides with a cellular model, while Hergarten et al. (2000) and Hergarten and Neugebauer (2001) carried out a similar exercise for drainage networks.

Modelled and observed drainage basins and their channel networks were also the focus of the work of Rodríguez-Iturbe and co-workers (e.g. Rodríguez-Iturbe and Rinaldo, 1997). This investigation concentrated on the fractal and multifractal properties of channel networks. The models developed by Rodríguez-Iturbe and co-workers are based on the continuity equations for flow and sediment transport, which are solved on a two-dimensional, rectangular grid. A similar approach is used in the long-term landscape evolution modelling of Coulthard et al. (2002), which focuses on small catchments, and that of Tucker and Slingerland (1996 and 1997) which has its emphasis on larger catchments.

De Boer (2001) used a CA approach to model the dynamics of sediment movement in fluvial drainage networks. This work is more fully described in 7.3.1 below.

The next generation

All modelling approaches from this first generation of studies have been 'conceptual' in the sense that arbitrary units are used for cell size, elevation etc. In many cases (Murray and Paola, 1997), the scaling of their model is undefined, so that the size of the landscape area being simulated is unclear. Thus validation of these models by comparison of model results with observations is impossible.

While these models have clearly established the general validity and potential of the CA approach for landscape modelling, an obvious next step is to use more realistic units and to construct the models in such a way that their outputs can be more rigorously validated. This is the aim of the RillGrow 2 model, as described in 7.3.2 below.

7.3 Applications of self-organizing systems concepts in geomorphology

7.3.1 Landscape-scale

Cascade 5 (De Boer, 2001) is a model of landscape evolution that can be used to illustrate and investigate many aspects of self-organization and complexity in fluvial landscapes.

7.3.1.1 Cascade 5: model description

As stated earlier, all models are to some extent based on simplifying reality. Cascade 5 is a highly simplified representation of a fluvial landscape. It does, however, capture the most essential features and processes. In its present form, Cascade 5 does not incorporate weathering, soil formation, sub-surface flow, large mass movements, or a distinction between channel and slope processes. As with most CA models, adjacent cells in a two-dimensional grid interact according to a set of rules. The value in each grid cell represents the local elevation. Results presented here were derived using a 210 by 210 elevation grid (i.e. a DEM: Digital Elevation Model) for a total of 44,100 cells.

Each model run starts with an initial topography which is then modified during the simulation. The results of the model runs discussed here were obtained with an initial topography with a random elevation for each cell in arbitrary units; this ranged between 0 and 1. To create relief in the model, all cells on the edge of the grid had their elevation lowered by 100, and base level was set at the elevation of the lowest cell on the grid's edge. Earlier investigations showed that, given enough time, the model modifies the initial topography to such an extent that the starting point for long model runs is, in the end, immaterial.

7.3.1.2 Transport laws in Cascade 5

Following Kirkby (1976), Armstrong (1976 and 1987) and others, sediment fluxes in Cascade 5 are calculated with the equation:

$$F = a \cdot S^b + c \cdot \sin(\arctan(S)) \tag{1}$$

where:

F is the total sediment flux from the originating cell to its lowest neighbour

S is the elevation difference between the originating cell and its lowest neighbour

a, b and c are constants.

In Equation (1), $a \cdot S^b$ is the wash component, the sediment flux caused by flowing water; and $c \cdot \sin(\arctan(S))$ is the creep component, the sediment flux caused by small mass movements. All results described herein were obtained using a = 0.015, b = 1.5 and c = 0.01. With b>1, the wash component increases faster than the elevation difference between cells, and at high values of the elevation difference, the wash component becomes the dominant part of the total sediment flux (Figure 7.2). Conversely, for small values of the elevation difference, small mass movements (i.e. creep) become the dominant sediment transfer process between cells (Figure 7.2).

Values of a, b and c in Equation (1) were selected such that the maximum sediment flux from an originating cell to its lowest neighbour is less than half the elevation difference between the two cells. Thus the lowest neighbour cannot become higher than the originating cell in one time step.

It is worth noting that the present form of Equation (1) does not incorporate a threshold elevation difference or slope below which sediment transport ceases. The reason that a threshold slope was not used is that the choice of its value would have a significant impact on the morphology and structure of the modelled landscape. No threshold slope is used in the model to avoid dealing with an additional parameter, and to enable the long-term evolution of the landscape towards a peneplain.

At the grid's edge, sediment is exported from the grid, and sediment yield is calculated using a modified form of Equation (1):

$$F = a \cdot S_B^b + c \cdot \sin(\arctan(S_B)) \tag{2}$$

where S_B is the elevation difference between the cell on the edge and base level. Sediment yields indicate relative differences in sediment yield through time and between cells on the edge of the grid.

7.3.1.3 Model rules

The Cascade 5 model operates by applying the rules in Table 7.2. Starting with the initial topography, the model is allowed to run until all traces of the original topography are removed and there are no more dramatic changes in the topography. During the model runs, a detailed record is kept of precipitation (storm area and location) and sediment export from the grid (sediment yield, coordinates of cells from which export occurs and base level). In addition, the elevations at all grid cells are recorded periodically for analyses of topography, drainage network configuration and mass balances.

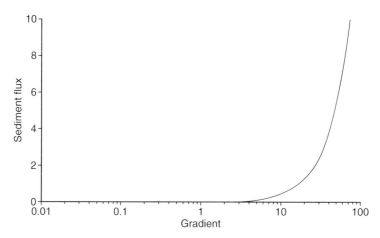

Figure 7.2 *Cascade 5: Relationship between gradient and erosion rate at the grid cell scale (Equation 1).*

7.3.1.4 Base level dynamics

Base level plays an important role in the model since it determines the rate of lowering of the cells on the grid's edge. If base level is held constant, and in the absence of a threshold slope, all cells will be worn down gradually until, ultimately, the elevation in all cells equals base level. In such a configuration, the model simulates the landscape as a closed system, which receives an initial input of potential energy through uplift, after which the system runs down to the high entropy state of, in Davis' terms, a peneplain (Chorley et al., 1984). Much more interesting, however, is the study of landscapes that function as dissipative systems, the class of open systems in which a throughput of energy and matter results in complex behaviour and the formation of ordered structures within the system. In Cascade 5, a periodic increase in potential energy is obtained by lowering the base level each time material is exported from one of the cells on the edge of the grid. Lowering base level in response to sediment export can be interpreted as incorporating a relative base level drop resulting from instantaneous isostatic rebound. In the model, base level was lowered because this only involves changing one variable every time sediment is exported from the grid. The

Table 7.2 *The rules of the Cascade 5 model*

Rule	Description
1	Randomly select the size of the area affected by the rainstorm. Rainstorms in the model are square, and can range in size from 1×1 to 210×210. For this paper, the probability of occurrence of a rainstorm is inversely proportional to its areal extent so that, for example, the probability of occurrence of a rainstorm covering 3×3 cells is 1/9 of the probability of occurrence of a rainstorm covering 1 cell. This results in a rainstorm frequency distribution given by $f \propto A^{-1}$ where f is the frequency of occurrence and A is the storm area.
2	Randomly select the location of the rainstorm. The location of each rainstorm is selected so that the entire rainstorm occurs within the 210×210 grid.
3	Make a list of all the cells receiving precipitation. All cells in the area of the rainstorm receive one unit of precipitation.
4	Process all cells in the list once, following rules 5, 6 and 7.
5	If the cell is on the edge of the grid, its elevation is lowered by an amount calculated with Equation (2) and the cell is removed from the list.
6	If the cell has no neighbour with an elevation lower than its own, it is a depression in which sediment is deposited and the cell is removed from the list.
7	Find the originating cell's lowest neighbour. Lower the elevation of the originating cell by an amount calculated using the transport law in Equation (1) and increase the elevation of its lowest neighbour, the receiving cell, by this amount. In the list, replace the coordinates of the originating cell by the coordinates of the receiving cell.
8	Go back to rule 4 until the list is empty.
9	Go back to rule 1 to start a new rainstorm.

alternative method, increasing the elevation of all cells, is computationally far less effective as it involves changing 44,100 elevation values every time sediment export takes place.

7.3.1.5 Spatial and temporal scales of Cascade 5

The objective of Cascade 5 is to evaluate how local rules of sediment transport result in global patterns in the sediment dynamics, rather than to model the evolution of a specific landscape to predict or postdict its sediment transport record. As a result, the spatial and temporal scales of the model need not be strictly defined, other than that, in general terms, the model concerns large temporal and spatial scales. An important spatial scale consideration is that each cell represents an area with a size of the order of 10^1 to 10^2 square kilometres. Thus, the processes in each cell are a combination of channel and slope processes, and no distinction is made between channel and slope cells. In terms of temporal scale, the model is aimed at evaluating landscape evolution and sediment dynamics over periods of 10^3 to 10^5 years.

7.3.1.6 Results of Cascade 5 model runs

A landscape in Cascade 5 forms a dissipative system in which individual cells interact with their neighbours. Groups of cells form drainage basins, which again interact with adjacent, larger drainage basins and in turn form part of larger drainage basins, which also interact with adjacent drainage basins and are part of still larger drainage basins and so on. In this way, the repeated application of the local rules describing the interaction between adjacent cells results in a hierarchical structure displaying global properties. The rules describing these global properties, however, cannot be deduced from the local rules. Some global properties characterize the morphology such as the frequency distribution of slope angles or the Horton ratios that summarize the drainage network configuration. Other global properties describe how the landscape functions as a unit and enable a process-based description at the global scale.

Equation (1) is the local rule for sediment transport in Cascade 5 and relates sediment flux and gradient. At the landscape scale, the simple relationship between gradient and sediment flux no longer applies. A plot of average gradient and erosion rate in a grid cell indicates that the relationship between these two variables at the landscape scale displays substantial scatter (Figure 7.3), and it must be concluded that, at the landscape scale, gradient is not the controlling variable. Instead, the location of a cell in the landscape determines the erosion rate. The fluvial landscape that evolves in Cascade 5 is, of course, organized into drainage basins. Within a drainage basin, there is a strong relationship between elevation and erosion rate (Figure 7.4), especially in the lower reaches of the drainage system. Figure 7.4a shows the relationship between elevation and erosion rate for all cells in a drainage basin. The scatter in the data is greatly reduced by selecting just those cells lower in the drainage system, for example, cells receiving sediment from 25 or more other cells (Figure 7.4b). The scatter is reduced because lower in the drainage network there is less chance of finding cells that have been

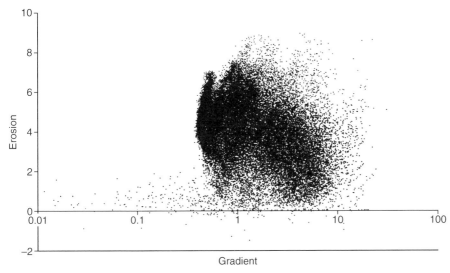

Figure 7.3 *Relationship between gradient and erosion rate at the landscape scale.*

subjected to the effects of fewer than average rainstorms. The relationship between elevation and erosion rate is an example of a global property of the landscape that results from the repeated application of the local rules, but cannot be predicted from the local rules.

In the case of Cascade 5, sediment transport at the local scale is given by Equations (1) and (2). The global rules describing sediment transport at the landscape scale can be formulated in various ways. An important functional characteristic of a fluvial landscape is a measure of the total amount of sediment removed from the landscape over a certain time interval. The frequency distribution of the total sediment yield is the sum of the contributions of all rainstorms of all sizes. Figure 7.5a shows the frequency distributions of total sediment yield for storms with areas of one, four, nine, sixteen and twenty-five cells, respectively. As a group, storms of nine cells contribute less than storms of four cells, which in turn contribute less than the storms of one cell. The inverse relationship between storm area and the total contribution to the total sediment yield is the result of the frequency distribution of the rainstorms, according to which the frequency of occurrence of a rainstorm is inversely proportional to its area. Normalized frequency distributions of the total sediment yield for storms of a specific magnitude can be obtained by scaling the frequency distribution of the sediment yield with rainstorm magnitude and frequency (Figure 7.5b). From Figure 7.5b it is evident that the sediment yield of a rainstorm can range over more than three orders of magnitude, depending on the storm location and the recent history of the landscape. The normalized frequency distributions in Figure 7.4 are a property of the entire landscape and cannot be predicted from the small-scale sediment transport rules of

Equation (1), even though these distributions result from the repeated application of these rules. The frequency distributions of sediment yield, therefore, are an emergent property of the landscape.

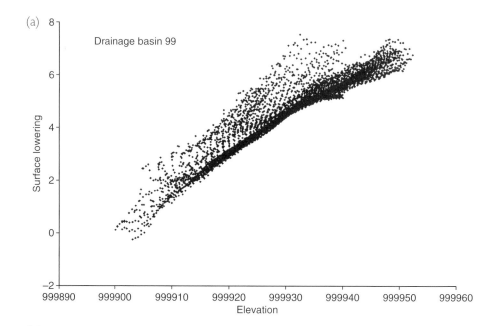

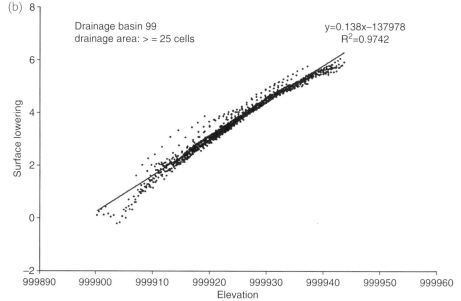

Figure 7.4 *Cascade 5: Relationship between elevation and erosion rate*
(a) for all grid cells in drainage basin 99 and
(b) for all grid cells in drainage basin 99 with a drainage area >= 25 cells.

7.3.2 Plot-scale

7.3.2.1 The RillGrow I model

Favis-Mortlock (1996 and 1998b) constructed the RillGrow I model in order to test the hypothesis that hillslope rill initiation and development may be modelled using a self-organizing systems approach, i.e. driven by simple rules governing systemic interactions on a much smaller scale than that of the rills. The central idea was that some rills are more 'successful' than others, since they preferentially grow in size and sustain flow throughout a rainfall event. Thus they compete for runoff. From an initial population of many microrills, only a subset subsequently develops into larger rills as part of a connected network (Table 7.3).

Erosive modification of the soil's microtopography produces a positive feedback loop, with the most 'successful' rills (i.e. those conveying the most runoff) modifying the local microtopography to the greatest extent, and so most effectively amplifying their chances of capturing and conveying future runoff. There is a limit to this growth, however (i.e. an associated negative feedback); each 'successful' rill's catchment cannot grow forever, since eventually the whole soil surface will be partitioned between the catchments of the 'successful' rills. The dynamics of this competitive process give rise to connected rill networks. Thus the hillslope erosional system is a dissipative system, with rainfall providing the essential input of matter and energy to the system, and runoff and sediment being the outputs.

The very simple RillGrow I CA model produced apparently realistic results (Favis-Mortlock et al., 1998) in terms of reproducing the observed characteristics of rill networks on plot-sized areas. However, it possessed some serious conceptual limitations which prevented the approach from being more rigorously validated. Firstly, the algorithm used (broadly similar to the precipiton approach of Chase (1992) described above) meant that the model did not operate within a true time domain, and so validation of relationships with a temporal aspect (e.g. the effects of rainfall intensity or time-varying discharge) was impossible. In addition, the model assumed an infinite transport capacity, with erosion being entirely detachment-limited. Thus the model could only hope to correctly reproduce situations where deposition is minimal. Finally, the model possesses a rather weak physical basis.

Table 7.3 'Successful' and 'unsuccessful' rills. From Favis-Mortlock (1996).

Category of rill	Rate of growth	Effectiveness during rainfall event
Successful	Higher	Becomes major carrier for runoff and eroded soil for part of hillslope; may 'capture' weaker rills
Unsuccessful	Lower	Becomes less and less important as a carrier for runoff and sediment; may eventually be 'captured' or become completely inactive

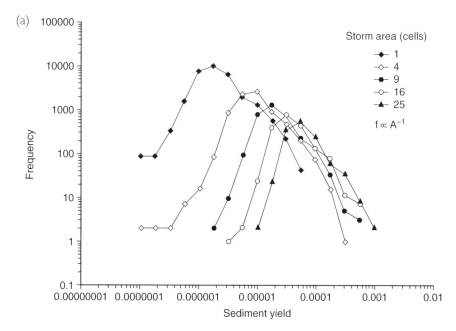

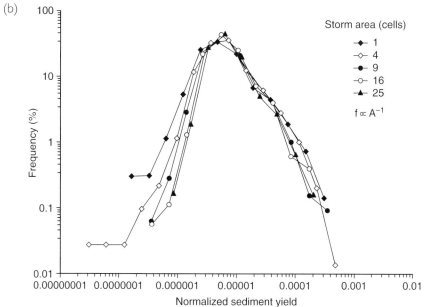

Figure 7.5 *(a) Cascade 5: frequency distributions of total sediment yield for storms of 1, 4, 9, 16 and 25 cells in area*
(b) Cascade 5: normalized frequency distribution of total sediment yield for storms of 1, 4, 9, 16 and 25 cells in area.

In many respects, RillGrow 1 was a typical 'first-generation' geomorphological CA model. In order to move beyond these limitations, RillGrow 2 was developed, the aim being to improve the process descriptions of the first version of the model, while (as far as possible) still retaining its simplicity. An early version of RillGrow 2 was described in Favis-Mortlock et al. (2000). The most recent version is summarized here. A detailed description of the model is forthcoming (Favis-Mortlock, in preparation).

7.3.2.2 The RillGrow 2 model

RillGrow 2, like RillGrow 1, operates upon an area of bare soil which is specified as a grid of microtopographic elevations (a DEM). Typically, cell size is a few millimetres, with elevation data either derived from real soil surfaces (Figure 7.6) by using a laser scanner (Huang and Bradford, 1992) or by means of photogrammetry (Lascelles et al., 2002); or generated using some random function (cf. Favis-Mortlock, 1998b). Computational constraints mean that, for practical purposes, the microtopographic grid can be no larger than plot-sized. A gradient is usually imposed on this grid.

The model operates using a fixed timestep, which is typically of the order of 0.05 sec. At each time step, multiple raindrops are dropped at random locations on the grid, with the number of drops depending on rainfall intensity. Run-on from upslope may also be added at an edge of the grid. Usually the soil is assumed to be saturated so that no infiltration occurs; however, a fraction of all newly arrived water may be removed if necessary, as a crude representation of infiltration losses.

Splash redistribution is simulated in RillGrow 2. Since this is a relatively slow process it is not normally calculated every time step. The relationship by Planchon et al. (2000) is used: this is essentially a diffusion equation based on the Laplacian, with a 'splash efficiency'

Figure 7.6 Soil surface microtopography: the scale at which RillGrow 2's rules operate. The finger indicates where flow is just beginning to incise a microrill in a field experiment (see Lascelles et al., 2000). Photograph © Martin Barfoot, 1997 (martin.barfoot@geog.ox.ac.uk). Used by permission.

term which is a function of rainfall intensity and water depth. Currently, the splash redistribution and overland flow components of RillGrow 2 are not tightly coupled; in particular, none of the sediment moved by splash is added to the overland flow.

Movement of overland flow between 'wet' cells occurs in discrete steps between cells of this grid. Conceptually, overland flow in RillGrow 2 is therefore a kind of discretized fluid rather like the 'sheepflow' illustrated in Figure 7.7.

For the duration of the simulation, each 'wet' cell is processed in a random sequence which varies at each time step. The logic outlined in Figure 7.8 is used for the processing.

Outflow may occur from a 'wet' cell to any of the eight adjacent cells. If outflow is possible, the direction with the steepest energy gradient (i.e. maximum difference in water-surface elevation) is chosen. The potential velocity of this outflow is calculated as a function of water depth and the velocity of adjacent 'wet' cells, if any (i.e. the vertical and horizontal components of hydraulic radius). However, outflow only occurs if sufficient time has elapsed for the water to cross this cell. Thus outflow only occurs for a subset of 'wet' cells at each time step.

When outflow does occur, the transport capacity of the flow is calculated using the previously calculated flow velocity with this S-curve relationship (Equation (5) in Nearing et al., 1997):

$$\log_e(q_s) = \frac{\alpha + \beta \cdot e^{\gamma + \delta \log_e(\varpi)}}{1 + e^{\gamma + \delta \log_e(\varpi)}} \tag{2}$$

Figure 7.7 *'Sheepflow': a visual analogy of RillGrow 2's discretized representation of overland flow. Photograph © Martin Price, 1999 (martin.price@perth.uhi.ac.uk). Used by permission.*

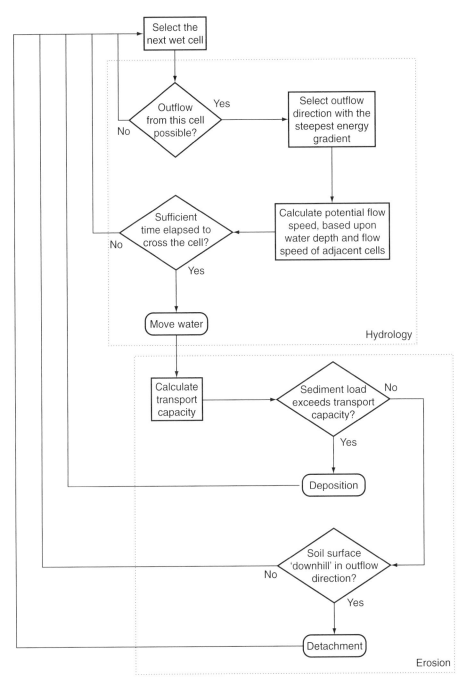

Figure 7.8 *A partial flowchart of the RillGrow 2 model.*

where:

q_s is unit sediment load

$\alpha, \beta, \gamma, \delta$ are constants

and:

$$\varpi = \rho \cdot g \cdot S \cdot q$$

where:

ω = stream power

ρ = density of water

g = gravitational acceleration

S = energy slope

q = unit discharge of water

If the sediment concentration of the water on the cell exceeds its transport capacity, deposition occurs. This is calculated using a version of equation 12 in Lei et al. (1998), which assumes deposition to be a linear function of the difference between sediment load and transport capacity. Note that no consideration is made of different settling times for each size fraction of the deposited sediment, or of differences in properties (e.g. bulk density) between re-entrained and newly eroded sediment.

Whereas outflow velocity and transport capacity are both determined by the energy gradient, it is assumed to be the soil surface gradient which controls detachment. In RillGrow 2, detachment occurs only if both energy gradient and soil surface gradient are downhill. If the soil surface gradient is uphill, then the energy gradient-driven outflow is assumed to be merely hydrostatic levelling. Where detachment does occur, its value is calculated using a probabilistic equation by Nearing (1991). This assumes that detachment is controlled by the probability of occurrence of random turbulent bursts and the difference between soil strength and shear stress. The relationship used in RillGrow 2 is a reformulation of Equation (10) in Nearing (1991):

$$e = K \cdot S \cdot u \cdot P \tag{4}$$

where:

e = detachment

K = a constant

u = outflow speed

and:

$$P = 1 - \Phi \, \frac{T - \tau_b}{\sqrt{S_T^2 + S_{\tau_b}^2}}$$

where:

Φ = the cumulative probability function of a standard normal deviate

T = a constant

S_T = the coefficient of variation of T (assumed constant)

$S_{\tau b}$ = the coefficient of variation of $?_b$ (assumed constant)

and:

$$\tau_b = 150 \cdot \pi \cdot g \cdot h \cdot S$$

where:

ρ = density of water

g = gravitational acceleration

h = water depth

S = energy slope.

RillGrow 2, while still relatively simple, is thus a development of the earlier RillGrow model in that it attempts to explicitly reproduce, in a true time domain, the effects of several processes involved in rill formation.

7.3.2.3 Results from RillGrow 2

During a series of 12 laboratory-based experiments at the University of Leicester (Lascelles et al., 2000, 2002 and Favis-Mortlock et al., in preparation) simulated rainfall (i.e. from an overhead sprinkler system) was applied to sandy soil in a 4 × 1.75m flume. A range of slope angles and rainfall intensities were used (Table 7.4). Each experiment lasted for 30 minutes, during which surface flow, discharge, sediment removal and flow velocities were measured. Prior to and following each experimental run, digital photogrammetry was used to create DEMs of the soil's surface. The initial (pre-experiment) DEMs and other data were then used as inputs to RillGrow 2. For one experiment only (X11: Table 7.4), measured flow velocities, discharge and sediment yield were also used to calibrate the model with respect to soil roughness and erodibility. Once calibrated, the model's inputs were not further adjusted when simulating the other 11 experiments.

Full results are given in Favis-Mortlock et al. (in preparation); however, measured and simulated discharge and sediment delivery for all experiments are compared in Figure 7.9. While both total discharge and sediment loss were well simulated in all cases, sediment loss was consistently underestimated for the low-gradient experiments. This is thought to be because of the loose coupling between the splash and flow components of the model: splashed sediment is not added to the flow, hence it is not lost at the flume end. While this underestimation occurs at all gradients, it is most noticeable when flow detachment is low.

A more difficult test is for the model to reproduce the observed patterns of erosion at the end of the experiments. Results for two experiments are shown in Figure 7.10.

For both experiments, the model was able to reproduce the main elements of the rill pattern; again, more successfully for the high-gradient case (X11) than the low-gradient one (X12). A similar result was seen for most of the other 10 experiments. Thus, with

Table 7.4 *Some characteristics of the Leicester flume experiments used to evaluate RillGrow 2. From Favis-Mortlock et al. (in preparation).*

Experiment No.	Slope angle (degrees)	Rainfall (mm hr⁻¹)
X09	5	108
X10	10	129
X11	15	125
X12	5	117
X13	10	127
X14	15	126
X15	5	120
X16	10	126
X17	15	131
X18	5	121
X19	10	123
X20	15	125

microtopography as the only spatially explicit input, RillGrow 2 is able to predict the pattern of rill networks which will be formed. It does this by considering rill networks to be emergent, whole-system responses to interactions between flow, detachment and deposition at a scale of millimetres.

7.4 Discussion: how does all this help?

We started with a question about the explanatory power of geomorphology with respect to features of real landscapes. En route, we looked at:

- landscape models: in particular, their apparently ever-increasing complexity
- recent advances from a variety of sciences in understanding non-linear systems, starting with deterministic chaos and followed by self-organization and emergence.

We have also glanced at the inroads which these ideas have made within physical geography, and looked at two geographical models which employ these notions. But how does this help us to answer the original question?

7.4.1 Does complexity mean complicated?

The evidence does indeed seem to support the idea that complex landscapes are not necessarily the result of complex processes. Systems which self-organize need only be simple at the 'component' level, yet are capable of producing a great wealth of complex, emergent responses. Thus, simple models such as the two described here are capable of producing realistic facsimiles of some of the complex features of real landscapes, and we suggest here that the mounting weight of model-based evidence, together with the

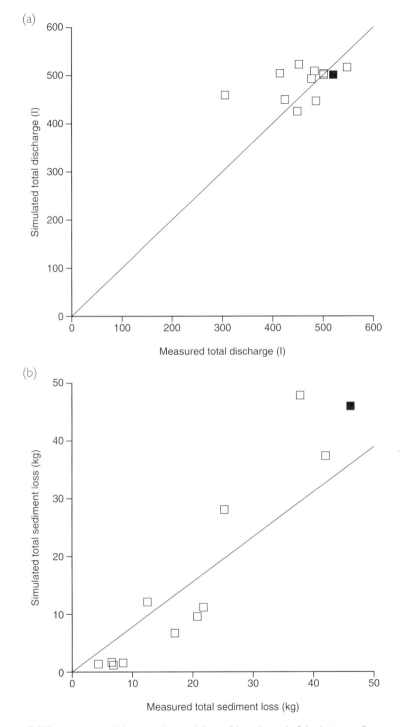

Figure 7.9 *Total discharge (a) and sediment delivery (b) at the end of the Leicester flume experiments, as measured and simulated by RillGrow 2. The black square indicates the value for X11; this was used to calibrate the model. From Favis-Mortlock et al. (in preparation).*

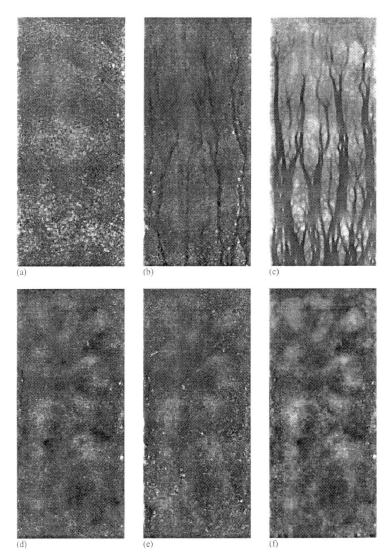

Figure 7.10 *Photogrammetry was used to derive DEMs of the soil's surface in the Leicester flume, shown here in plan view. In each case, darker areas have lower elevation. (a) X11 initial (pre-experiment), (b) X11 final (post-experiment), (c) X11 simulated by RillGrow 2; (d) X12 initial, (e) X12 final, (f) X12 simulated. From Favis-Mortlock et al. (in preparation).*

field- and laboratory-based studies listed in 7.2.3.1 above, tends to confirm our belief that self-organization is a real feature of real landscapes.

What we do not know, and perhaps can never know, is the extent to which such models capture the fundamentals, the underlying essences of the process relationships which create real landscapes. Because of the spectre of 'model equifinality' (Beven, 1989) it is possible that even the most perfect resemblance between model and reality is merely due to chance.[28] Oreskes et al. (1994, 641) went so far as to state that 'verification and validation of numerical models of natural systems is impossible', and argued that 'the primary value of models is heuristic.' Following their argument, a resemblance between model and reality is not a test of a model.

7.4.1.1 A route back to simplicity for landscape models?

If self-organization is a real feature of real landscapes, then one implication is a message of hope for landscape modellers. We are not doomed to ever more complex models! While it may seem perverse that the study of complexity can lead us to simplicity, ground-breaking research on self-organization during the late 1980s and 1990s appears to have been a fresh wind, doing away with the tired idea of fitting the results of data-hungry models to sparse, observed data. Instead of the cumbersome drudgery of varying a large number of parameters and variables to obtain a better fit between the computed and observed data, models may again be used to generate new and exciting ideas.

This is a dramatic shift in modelling philosophy. It comes about because models based on self-organization have the potential to develop 'new' knowledge about the systems they represent.[29] Furthermore, this shift provides an opportunity to re-evaluate the familiar conceptual tools in the geomorphological toolkit, e.g. to lead us to a new view of the function of positive and negative feedback in geomorphology, and to a change in appreciation of the equilibrium concept (Thorn and Welford, 1994a; Kennedy, 1994; Phillips and Gomez, 1994; Thorn and Welford, 1994b; and Clifford, 2001).

7.4.1.2 Of polygenesis and palimpsests

Of course, many (most?) landscapes are palimpsests, with the features formed during earlier periods surviving into later periods as relicts. Thus much of northern Europe and the USA is a relict periglacial landscape, for example. Such relict features may be reactivated if conditions should revert.

While this complicates our main argument – that complex landscapes are the result of simple small-scale process interactions, and may be modelled using simple rules – it does not invalidate it. For such polygenetic landscapes, the configuration achieved at the end of one phase forms the boundary conditions for the next. Thus in modelling such landscapes, the way to proceed would be to treat each landscape-forming phase separately, and to 'chain' them so that the outputs of one phase (a given landscape configuration) become the inputs of the next. This general approach was adopted in a modelling study of soil loss during the whole period of agriculture on the UK South Downs (Favis-Mortlock et al., 1997), albeit for one of the simplest spatial situations: a single hillslope field.

---

28 At the deepest level, why should mathematics be of any value at all in describing the world? See Wigner, 1960.
29 See footnote 25.

7.4.1.3 But is it really back to simplicity?

In the distinctly post-modern spirit of 'there's no such thing as a free lunch', we must ask ourselves: does this approach *really* represent a return to simplicity? For example, whereas the RillGrow 2 model has relatively modest data requirements, a very high DEM resolution is essential (Groundwater,[30] 2002; cf. Schoorl et al., 2000), and computer-processing requirements are daunting for large grids. Is it possible that we are just trading one set of data problems for another?

There are other possible hidden complications. For example, what are the implications of self-organization for the application of the ergodic hypothesis? This is an assumption which is widely used in geomorphological studies: since change in landscapes occurs over timescales longer than human observation, observation of landforms at different locations which are at different stages of their evolution allows us to infer the sequence of change, and hence the slow-acting processes which produce that change. Thus the ergodic principle, which states that space may be substituted for time (or vice-versa), is invoked (Craig, 1982; Culling, 1987 and Favis-Mortlock et al., 2001). Yet implicit in this assumption is that landscape features are essentially interchangeable: thus in a practical sense, the ergodic assumption is more likely to be valid for small-scale features (which are likely to be numerous) than for large-scale features (since there will probably be too few large-scale features for all realizations to have taken place). As a result, large-scale systems tend to display historical path dependence, and investigations of large-scale systems are typically dominated by historical aspects.

Self-organization, with its increased emphasis on the temporal and spatial context in which landscape features occur (see 7.2.2.5 above), may demand a rethink of our usage of ergodicity in geomorphology. An increased contextual component suggests an increased uniqueness of individual landscape features[31] irrespective of scale. This is not a new insight: Schumm (1991, 45) says that, 'The more specific our questions, the shorter the timespan with which we must deal'. But, if we acknowledge the role of self-organization in shaping landscapes, it is an example of the kind of rethinking which this change of perspective may force upon us.

7.4.1.4 Why expect simplicity anyway?

A central premise of this chapter is of some underlying simplicity of landscape-forming processes. But why should we expect things to be simple? 'Occam's Razor'[32] is usually invoked here, but this is a metaphysical guideline. While Isaac Newton agreed ('We are to admit no more causes of natural things than such as are both true and sufficient to

---

30 This study makes use of the RillGrow 1 model, but results from a similar analysis of RillGrow 2 would probably be substantially similar.
31 Bar-Yam (1997, 94), from the more rigorous physics-based perspective of ergodicity (Favis-Mortlock et al., 2001) points out that 'Any system that breaks symmetry violates the ergodic theorem'.
32 'Entities should not be multiplied unnecessarily', William of Ockham (1285–1347?), *Quodlibeta*. Actually this is something of an Anglo-Saxon notion, since Aristotle had earlier discussed the desirability of making the simplest choice between conflicting explanations, (David Ruelle, personal communication, 2000).

explain their appearances ... Nature is pleased with simplicity, and affects not the pomp of the superfluous causes'. Newton, 1729), other more recent thinkers have had their doubts. Albert Einstein pointed out that 'the most incomprehensible thing about the world is that it is comprehensible'.[33] In other words, why should we expect the abilities of humanity to comprehend and describe natural processes to be adequate to do so, even partially? The physicist Richard Feynman (1999, 186), however, saw even less merit in Occam's Razor as an ultimate guiding principle:

> I know that there are some scientists who go about preaching that Nature always takes on the simplest solutions. Yet the simplest solution by far would be nothing, so that there should be nothing at all in the universe. Nature is far more inventive than that, so I refuse to go along thinking it always has to be simple.

So is this desire for simplicity merely wishful thinking? A full discussion is clearly infeasible here. However, a thought-provoking observation was made by physicist Werner Heisenberg: 'What we observe is not nature itself, but nature exposed to our method of questioning' (Heisenberg, 1958). Could it be that both simplicity and complexity are, to an extent, of our own making? According to Poincaré (1908): '... by natural selection our mind has adapted itself to the conditions of the external world. It has adopted the geometry most advantageous to the species or, in other words, the most convenient. Geometry is not true, it is advantageous'. Something similar may be true of all mathematical description (cf. Wigner, 1960).

Perhaps the last word should come from mathematician, Ian Stewart: here is a section from his recent novel, *Flatterland,* which describes attempts by humans ('Peoples') to understand the world mathematically (Stewart, 2001, 268):

> ... Some of them even realised that what they would get would be a description of how the universe seems to work, not the actual rules by which it runs. Because – well, because there might not be any such rules. A rule is a very Peopleish concept, and even more so a law. Those are methods Peoples use to run their own society. So it seems to me that they had a mental picture of their universe that was modelled on their own social interactions. The amazing thing was that it worked pretty well.

33 Actually a paraphrase (Calaprice, 2000, 278). Einstein also noted though that 'A theory is the more impressive the greater the simplicity of its premises, the more different kinds of things it relates, and the more extended its area of applicability' (Calaprice, 2000, 262). Much the same thing is said by Bar-Yam (1997, 257), 'We have argued that the purpose of knowledge is to succinctly summarise information that can be used for prediction. Thus, in its most abstract form, the problem of deduction or prediction is a problem in data compression. It can thus be argued that science is an exercise in data compression. This is the essence of the principle of Occam's Razor and the importance of simplicity and universality in science. The more universal and the more general a law is, then the more data compression has been achieved'.

7.5 Conclusions

Through the years, physical geography has gone through a series of distinct phases, each characterized by a particular set of paradigms and tools that determine the direction of research, the kind of questions posed, and the nature of the answers found. General systems concepts were introduced to physical geography by Chorley (1962) and Chorley and Kennedy (1971). The view of the physical environment as a series of linked and nested systems led to an increased effort to measure process rates and, coupled with the technological advances and increased availability of computers, lead to the rapid increase in the development of computer models in physical geography. Notions of complexity and self-organization may characterize the next phase.

Returning to the question that began this chapter: perhaps it is time for geomorphology to mature as a science. Perhaps it is time to recognize that, apart from idle curiosity, explanation of specific features in the landscape is, by itself, of limited value. Where such explanations are essential is in validating our general models of landscape development. Thus it is of limited interest to know why the sediment load of the Rhine River varies as it does; but it is a different matter if knowledge of the Rhine's sediment-carry characteristics helps us to understand the unifying principles controlling the sediment loads of the River Exe, the Rhine River and the Amazon River. Similarly, we do not really need to explain the details of the drainage pattern of the South Saskatchewan River. Instead, we should look for similarities between, and common causes of, the drainage patterns of this river and other rivers at various scales in different regions. If such a shift in research emphasis from the primarily phenomenal (investigating what is found where) to the primarily intellectual (investigating how and why) is indeed our route-map for geomorphology in the 21st century, then an appreciation of the role of self-organization in creating the general patterns of landscapes is essential.

We finish with two quotes. The first concerns the importance of context in geomorphology (something that is vital to consider when landscapes are viewed from the perspective of self-organization – see 7.2.2.5 above), as appreciated a little under a century ago by William Morris Davis (1910, 75):

> ... the conviction dawns upon the learner that to attain even an elementary conception of what goes on in his own parish, he must know something about the universe; that the pebble he kicks aside would not be what it is and where it is unless a particular chapter of the earth's history ... had been exactly what it was ...

The second describes the self-organization of landscapes, as this would be perceived by a sufficiently long-lived observer:

> The hills are shadows, and they flow
> From form to form, and nothing stands;
> They melt like mist, the solid lands,
> Like clouds they shape themselves and go.
>
> Tennyson, A., *In Memoriam*, 1850

Acknowledgements

We thank John Boardman (University of Oxford) for comments on an earlier draft of this paper. DFM would like to thank the following for discussions which stimulated many of the ideas in this paper: John Boardman, Jack Cohen (University of Warwick), Anton Imeson (University of Amsterdam), Mark Nearing (USDA-ARS), John Thornes (King's College, London), Brian Whalley (Queen's University, Belfast); the 'self-organized lunchers' at Oxford (Susan Canney, Clive Hambler, Mike Packer and David Raubenheimer) and, finally, all those who attended the NATO-ASI 'Nonlinear Dynamics in Life and Social Sciences' (Moscow, 2000). Thanks also to the complex-science@necsi.org discussion list. Development of RillGrow 2 was partly funded by the UK Natural Environment Research Council (NERC GR3/11417: 'Rill initiation by overland flow and its role in modelling soil erosion').

References

Andrle, R. 1996: 'The west coast of Britain: statistical self-similarity vs. characteristic scales in the landscape'. *Earth Surface Processes and Landforms* 21(10), 955–962.

Aristotle c. 330 BCE: *Metaphysica*, 10f–1045a.

Armstrong, A.C. 1976: 'A three-dimensional simulation of slope forms'. *Zeitschrift für Geomorphologie, Supplementband* 25, 20–28.

Armstrong, A.C. 1987: 'Slopes, boundary conditions, and the development of convexo-concave forms – some numerical experiments'. *Earth Surface Processes and Landforms* 12(1), 17–30.

Bak, P. 1996: *How Nature Works*. Springer-Verlag, New York.

Bak, P., Tang, C. and Wiesenfeld, K. 1988: 'Self-organized criticality'. *Physical Review A* 38(1), 364–374.

Ball, P. 1999: *The Self-Made Tapestry: Pattern Formation in Nature*. Oxford University Press, Oxford.

Bar-Yam, Y. 1997: *Dynamics of Complex Systems*. Perseus, Reading, MA.

Berry, M. 1988: 'The electron at the end of the universe'. In L. Wolpert, and A. Richards, (eds.), *A Passion for Science*. Oxford Paperbacks, Oxford, 39–51.

Beven, K.J. 1989: 'Changing ideas in hydrology: the case of physically-based models'. *Journal of Hydrology* 105, 157–172.

Beven, K.J. 1996: 'The limits of splitting: hydrology', *The Science of the Total Environment* 183, 89–97.

Boardman, J. and Favis-Mortlock, D.T. 1999: 'Frequency-magnitude distributions for soil erosion, runoff and rainfall – a comparative analysis'. *Zeitschrift für Geomorphologie N.F. Supplementband* 115, 51–70.

Bohm, D. 1980: '*Wholeness and the Implicate Order*. Routledge, London.

Buchanan, M. 2000: *Ubiquity*. Weidenfeld and Nicolson, London.

Burges, A.E. 1938: *Soil Erosion Control (revised). A Practical Exposition of the New Science of Soil Conservation for Students, Farmers, and the General Public.* Turner E. Smith & Co., Atlanta GA.

Calaprice, A. (ed.), 2000: *The Expanded Quotable Einstein*. Princeton University Press, Princeton.

Çambel, A.B. 1993: *Applied Chaos Theory: A Paradigm for Complexity*. Academic Press, Boston.

Chaitin, G.J. 1999: *The Unknowable*. Springer-Verlag, Singapore.

Chase, C.G. 1992: 'Fluvial landsculpting and the fractal dimension of topography'. *Geomorphology* 5, 39–57.

Chorley, R.J. 1962: *Geomorphology and General Systems Theory*. US Geological Survey Professional Paper, No. 500–B.

Chorley, R.J. and Kennedy, B.A. 1971: *Physical Geography: A Systems Approach*. Prentice-Hall, London.

Chorley, R.J., Schumm, S.A. and Sugden, D.E. 1984: *Geomorphology*. Methuen, London.

Clifford, N.J. 2001: 'Physical Geography – the naughty world revisited'. *Transactions of the Institute of British Geographers* NS 26, 387–389.

Cohen, J. and Medley, G. 2000: *Stop Working and Start Thinking: A Guide to Becoming a Scientist*. Stanley Thornes, Cheltenham.

Cohen, J. and Stewart, I. 1994: *The Collapse of Chaos*. Penguin, London.

Coulthard, T.J., Macklin, M.G. and Kirkby, M.J. 2002: 'A cellular model of Holocene upland river basin and alluvial fan evolution'. *Earth Surface Processes and Landforms* 27(3), 269–288.

Coveney, P. and Highfield, R. 1995: *Frontiers of Complexity*. Faber and Faber, London.

Craig, R.G. 1982: 'The ergodic principle in erosion models'. In, C.E. Thorn, (ed.), *Space and Time in Geomorphology*. Allen & Unwin, Boston, 81–115.

Culling, W.E.H. 1987: 'Ergodicity, entropy and dimension in the soil covered landscape'. *Transactions of the Japanese Geomorphological Union* 8, 157–174.

Davies, G.L. 1969: *The Earth in Decay: A History of British Geomorphology 1578–1878*. Science History Publications, New York.

Davis, W.M. 1910: *Geographical Essays*. Dover Facsimile edited by D.W. Johnson, (1954), unabridged and unaltered republication of the work originally published by Ginn and Co., Dover, New York.

De Boer, D.H. 1992: 'Hierarchies and spatial scale in process geomorphology: a review'. *Geomorphology* 4, 303–318.

De Boer, D.H. 2001 'Self-organization in fluvial landscapes: sediment dynamics as an emergent property'. *Computers and Geosciences* 27(8), 995–1003.

Dickinson, W.T., Rudra, R.P. and Wall, G.J. 1986: 'Identification of soil erosion and fluvial sediment problems'. *Hydrological Processes* 1, 111–124.

Dikau, R. 1999: 'The need for field evidence in modelling landform evolution'. In S. Hergarten, and H.J. Neugebauer, (eds.), *Process Modelling and Landform Evolution*. Springer-Verlag, Berlin. 3–12.

Evans, I.S. and McClean, C.J. 1995: The land surface is not unifractal: variograms, cirque scale and allometry'. *Zeitschrift für Geomorphologie N.F. Supplement* 101, 127–147.

Favis-Mortlock, D.T. 1996: 'An evolutionary approach to the simulation of rill initiation and development'. In R.J. Abrahart, (ed.), *Proceedings of the First International Conference on GeoComputation (Volume 1)*. School of Geography, University of Leeds. 248–281.

Favis-Mortlock, D.T. 1998a: 'Validation of field-scale soil erosion models using common datasets'. In J. Boardman, and D.T. Favis-Mortlock, (eds.), *Modelling Soil Erosion by Water.* Springer-Verlag NATO-ASI Series I-55, Berlin, 89–128.

Favis-Mortlock, D.T. 1998b: 'A self-organizing dynamic systems approach to the simulation of rill initiation and development on hillslopes'. *Computers and Geosciences* 24(4), 353–372.

Favis-Mortlock, D.T. (in preparation): 'The RillGrow 2 model'.

Favis-Mortlock, D.T., Quinton, J.N. and Dickinson, W.T. 1996: 'The GCTE validation of soil erosion models for global change studies'. *Journal of Soil and Water Conservation* 51(5), 397–403.

Favis-Mortlock, D.T., Boardman, J. and Bell, M. 1997: 'Modelling long-term anthropogenic erosion of a loess cover: South Downs, UK'. *The Holocene* 7(1), 79–89.

Favis-Mortlock, D.T., Guerra, A.J.T. and Boardman, J. 1998: 'A self-organising dynamic systems approach to hillslope rill initiation and growth: model development and validation'. In W. Summer, E. Klaghofer, and W. Zhang, (eds.), *Modelling Soil Erosion, Sediment Transport and Closely Related Hydrological Processes.* IAHS Press Publication No. 249, Wallingford, UK, 53–61.

Favis-Mortlock, D.T., Boardman, J., Parsons, A.J. and Lascelles, B. 2000: 'Emergence and erosion: a model for rill initiation and development'. *Hydrological Processes* 14(11–12), 2173–2205.

Favis-Mortlock, D.T., Boardman, J. and MacMillan, V.J. 2001: 'The limits of erosion modeling: why we should proceed with care'. In R.S. Harmon and W.W. Doe III, (eds.), *Landscape Erosion and Evolution Modeling.* Kluwer Academic/Plenum Publishing, New York, 477–516.

Favis-Mortlock, D.T., Parsons, A.J, Boardman, J. and Lascelles, B. (in preparation): 'Evaluation of the RillGrow 2 model'.

Feynman, R.P. 1999: *Feynman Lectures on Gravitation.* Penguin, London.

Flake, G.W. 1998: *The Computational Beauty of Nature.* MIT Press, Cambridge, Massachusetts.

Gallagher, R. and Appenzeller, T. 1999: 'Beyond Reductionism'. *Science* 284, 79.

Garcia-Sanchez, L., Di Pietro, L. and Germann, P.F. 1996: 'Lattice-gas approach to surface runoff after rain'. *European Journal of Soil Science* 47(4), 453–462.

Gleick, J. 1987: *Chaos: Making a New Science.* Penguin, New York.

Goodwin, B. 1997: *How the Leopard Changed its Spots: the Evolution of Complexity.* Phoenix, London.

Greene, B. 1999 *The Elegant Universe:* Jonathan Cape, London.

Groundwater, P. 2002: *The Influence of Model Resolution on Rill Development: A Numerical Modelling Study.* Unpublished BSc Dissertation, School of Geography and the Environment, University of Oxford.

Haggett, P. and Chorley, R.J. 1967: 'Models, paradigms and the new geography'. In R.J. Chorley, and P. Haggett, (eds.), *Models in Geography.* Methuen, London, 19–41.

Hallet, B. 1990: 'Spatial self-organization in geomorphology: from periodic bedforms and patterned ground to scale-invariant topography'. *Earth Science Reviews* 29, 57–75.

Hansen, J.L., Van Hecke, M., Haaning, A., Ellegaard, C., Andersen, K.H., Bohr, T. and Sams, T. 2001: 'Instabilities in sand ripples'. *Nature* 410, 324.

Harvey, D. 1969: *Explanation in Geography,* Edward Arnold, London.

Hawking, S. 1988: *A Brief History of Time.* Bantam Books, New York.

Heisenberg, W. 1958: *Physics and Philosophy*, Harper & Row, New York.

Hergarten, S. and Neugebauer, H.J. 1998: 'Self-organized criticality in a landslide model'. *Geophysical Research Letters* 25(6), 801–804.

Hergarten, S. and Neugebauer, H.J. 1999: 'Self-organized criticality in landsliding processes'. In S. Hergarten, and H.J. Neugebauer, (eds.), *Process Modelling and Landform Evolution*. Springer, Berlin, 231–249.

Hergarten, S. and Neugebauer, H.J. 2001: 'Self-organized critical drainage networks'. *Physical Review Letters* 86(12), 2689–2692.

Hergarten, S., Paul, G. and Neugebauer, H.J. 2000: 'Modeling surface runoff'. In J. Schmidt, (ed.), *Soil Erosion: Application of Physically-Based Models*. Springer-Verlag, Berlin, 295–306.

Higgitt, D.L. and Rosser, N.J. 2002: 'Self-organisation in desert stone mantles'. Paper presented at RGS-IBG Session, '*Geomorphology: Chaos, Fractals and Self-organizing Systems*, January 2002, Belfast.

Holland, J.H. 1998: *Emergence: from Chaos to Order*, Perseus Books, Reading MA.

Huang, C. and Bradford, J.M. 1992: 'Applications of a laser scanner to quantify soil microtopography'. *Soil Science Society of America Journal* 56(1), 14–21.

Huggett, R.J. 1985: *Earth Surface Systems*. Springer-Verlag, Berlin.

Jakeman, A.J. and Hornberger, G.M. 1993: 'How much complexity is warranted in a rainfall–runoff model?' *Water Resources Research* 29, 2637–2649.

Jetten, V., de Roo, A.P.J. and Favis-Mortlock, D.T. 1999: 'Evaluation of field-scale and catchment-scale soil erosion models'. *Catena* 37(3/4), 521–541.

Jones, H. 1991: 'Fractals before Mandelbrot: a selective history'. In A.J. Crilly, R.A. Earnshaw, and H. Jones, (eds.), *Fractals and Chaos*. Springer-Verlag, Berlin, 7–33.

Kauffman, S. 1995: *At Home in the Universe: the Search for Laws of Self-Organization and Complexity*, Oxford University Press, Oxford.

Kennedy, B.A. 1994: 'Requiem for a dead concept'. *Annals of the Association of American Geographers* 84(4), 702–705.

Kirchner, J.W., Hooper, R.P., Kendall, C., Neal, C. and Leavesley, G. 1996: 'Testing and validating environmental models'. *The Science of the Total Environment* 183, 33–47.

Kirkby, M.J. 1976: 'Deterministic continuous slope models'. *Zeitschrift für Geomorphologie, Supplementband* 25, 1–19.

Kirkby, M.J., Naden, P.S., Burt, T.P. and Butcher, D.P. 1992: *Computer Simulation in Physical Geography* (Second Edition). John Wiley, Chichester.

Klemes, V. 1986: 'Dilettantism in hydrology: transition or destiny?' *Water Resources Research* 22, 177S–188S.

Klimontovich, Y.L. 2001: 'Entropy, information and ordering criteria in open systems'. In W. Sulis, and I. Trofimova, (eds.), *Nonlinear Dynamics in the Life and Social Sciences*. IOS Press, Amsterdam, 13–32.

Knisel, W.G. (ed.) 1980: *CREAMS – a Field Scale Model for Chemicals, Runoff and Erosion from Agricultural Management Systems*. US Department of Agriculture Research Report No. 26.

Laland, K.N., Odling-Smee, F.J. and Feldman, M.W. 1999: 'Evolutionary consequences of niche construction and their implications for ecology'. *Proceedings of the National Academy of Science* 96, 10242–10247.

Lascelles, B., Favis-Mortlock, D.T., Parsons, A.J. and Guerra, A.J.T. 2000: Spatial and temporal variation in two rainfall simulators: implications for spatially explicit rainfall simulation experiments'. *Earth Surface Processes and Landforms* 25(7), 709–721.

Lascelles, B., Favis-Mortlock, D.T., Parsons, A.J. and Boardman, J. 2002: 'Automated digital photogrammetry – a valuable tool for small-scale geomorphological research for the non-photogrammetrist?' *Transactions in GIS* 6(1), 5–15.

Lei, T., Nearing, M.A., Haghighi, K. and Bralts, V.F. 1998: 'Rill erosion and morphological evolution: a simulation model'. *Water Resources Research* 34(11), 3157–3168.

Lenton, T.M. 1998: 'Gaia and natural selection'. *Nature* 394, 439–447.

Leonard, R.A., Knisel, W.G. and Still, D.A. 1987: 'GLEAMS: Groundwater Loading Effects of Agricultural Management Systems'. *Transactions of the American Society of Agricultural Engineers* 30(5), 1403–1418.

Leopold, L.B., Wolman, M.G. and Miller, J.P. 1964: *Fluvial Processes in Geomorphology.* Freeman, San Francisco.

Lewin, R. 1997: 'Critical mass: complexity theory may explain big extinctions better than asteroids'. *New Scientist*, 23rd August 1997.

Link, W.K. 1954: 'Robot geology'. Cited on p21 of A.F. Pitty (1971): *Introduction to Geomorphology.* Methuen, London.

Mahnke, R. 1999: 'Pattern formation in cellular automaton models'. In J. Schmelzer, G. Röpke, and R. Mahnke, (eds.), *Aggregation Phenomena in Complex Systems.* Wiley-VCH, Weinheim, 146–173.

Mandelbrot, B.B. 1975: 'Stochastic models for the earth's relief, the shape and the fractal dimension of the coastlines, and the number-area rules for islands'. *Proceedings of the National Academy of Sciences, USA* 72(10), 3825–3828.

May, R.M. 1976: 'Simple mathematical models with very complicated dynamics'. *Nature* 261, 459–467.

Morgan, R.P.C., Quinton, J.N., Smith, R.E., Govers, G., Poesen, J.W.A., Chisci, G. and Torri, D. 1998: 'The EUROSEM Model'. In J. Boardman and D.T. Favis-Mortlock, (eds.), *Modelling Soil Erosion by Water.* Springer-Verlag NATO-ASI Series I-55, Berlin, 389–398.

Murray, A.B. and Paola, C. 1994: 'A cellular model of braided rivers'. *Nature* 371, 54–57.

Murray, A.B. and Paola, C. 1996: 'A new quantitative test of geomorphic models, applied to a model of braided streams'. *Water Resources Research* 32(8), 2579–2587.

Murray, A.B. and Paola, C. 1997: 'Properties of a cellular braided-stream model'. *Earth Surface Processes and Landforms* 22(11), 1001–1025.

Nearing , M.A. 1991: 'A probabilistic model of soil detachment by shallow turbulent flow'. *Transactions of the American Society of Agricultural Engineers* 34(1), 81–85.

Nearing, M.A., Foster, G.R., Lane, L.J. and Finkner, S.C. 1989: 'A process-based soil erosion model for USDA-Water Erosion Prediction Project technology'. *Transactions of the American Society of Agricultural Engineers* 32(5), 1587–1593.

Nearing, M.A., Norton, L.D., Bulgakov, D.A., Larionov, G.A., West, L.T. and Dontsova, K. 1997: 'Hydraulics and erosion in eroding rills'. *Water Resources Research* 33(4), 865–876.

Newton, I. 1729: *The Mathematical Principles of Natural Philosophy.* London. Translated by A. Motte (see http://www.fordham.edu/halsall/mod/newton-princ.html).

Nicolis, G. and Prigogine, I. 1989: *Exploring Complexity.* Freeman, New York.

Oreskes, N., Shrader-Frechette, K. and Belitz, K. 1994: 'Verification, validation, and confirmation of numerical models in the earth sciences'. *Science* 263, 641–646.

Phillips, J.D. 1995: 'Self-organization and landscape evolution'. *Progress in Physical Geography* 19, 309–321.

Phillips, J.D. 1996: 'Deterministic complexity, explanation, and predictability in geomorphic systems'. In B.L. Rhoads and C.F. Thorn (eds.), *The Scientific Nature of Geomorphology*. Wiley, New York, 315–335.

Phillips, J.D. 1997: 'Simplexity and the reinvention of equifinality'. *Geographical Analysis* 29(1), 1–15.

Phillips, J.D. 1999: 'Divergence, convergence, and self-organization in landscapes'. *Annals of the Association of American Geographers* 89, 466–488.

Phillips, J.D. and Gomez, B. 1994: 'In defense of logical sloth'. *Annals of the Association of American Geographers* 84(4), 697–701.

Pilotti, M. and Menduni, G. 1997: 'Application of lattice gas techniques to the study of sediment erosion and transport caused by laminar sheetflow'. *Earth Surface Processes and Landforms* 22(9), 885–893.

Pitty, A.F. 1971: *Introduction to Geomorphology*. Methuen, London.

Planchon O., Esteves M., Silvera N. and Lapetite J.M. 2000: 'Raindrop erosion of tillage induced microrelief. Possible use of the diffusion equation'. *Soil and Tillage Research* 56(3–4), 131–144.

Poincaré, J.H. 1908: *Science and Method*. 2001 (edn.), St. Augustine Press, Inc., South Bend, IN.

Portugali, J., Benenson, I. and Omer, I. 1997: 'Spatial cognitive dissonance and socio-spatial emergence in a self-organizing city'. *Environment and Planning B* 24, 263–285.

Richards, A., Phipps, P. and Lucas, N. 2000: 'Possible evidence for underlying non-linear dynamics in steep-faced glaciodeltaic progradational successions'. *Earth Surface Processes and Landforms* 25(11), 1181–1200.

Rodríguez-Iturbe, I. and Rinaldo, A. 1997: *Fractal River Basins: Chance and Self-organization*. Cambridge University Press, Cambridge.

Rosenblueth, A. and Wiener, N. 1945: 'The role of models in science', *Philosophy of Science* 12.

Ruelle, D. 1993: '*Chance and Chaos*'. Penguin, London. 195 pp.

Ruelle, D. 2001: 'Applications of chaos'. In, W. Sulis and I. Trofimova, (eds.), *Nonlinear Dynamics in the Life and Social Sciences*. IOS Press, Amsterdam, 3–12.

Samuel, E. 2002: 'What lies beneath?' *New Scientist,* 9th February 2002, 24–27.

Schoorl, J.M., Sonneveld, M.P.W. and Veldkamp, A. 2000: 'Three-dimensional landscape process modelling: the effect of DEM resolution'. *Earth Surface Processes and Landforms* 25(9), 1025–1034.

Schumm, S.A. 1991: *To Interpret the Earth: Ten Ways to be Wrong*. Cambridge University Press, Cambridge.

Sethna, J.P., Dahmen, K.A. and Myers, C.R. 2001: 'Crackling noise'. *Nature* 410, 242–250.

Smith, R. 1991: 'The application of cellular automata to the erosion of landforms'. *Earth Surface Processes and Landforms* 16, 273–281.

Solé, R.V., Manrubia, S.C., Benton, M., Kauffman, S. and Bak, P. 1999: 'Criticality and scaling in evolutionary ecology'. *Trends in Evolutionary Ecology* 14(4), 156–160.

Steegen, A., Govers, G., Nachtergaele, J., Takken, I., Beuselinck, L. and Poesen, J. 2000: 'Sediment export by water from an agricultural catchment in the Loam Belt of central Belgium'. *Geomorphology* 33, 25–36.

Stewart, I. 2001: *Flatterland,* Macmillan, London.

Takken, I., Beuselinck, L., Nachtergaele, J., Govers, G., Poesen, J. and Degraer, G. 1999

'Spatial evaluation of a physically-based distributed erosion model (LISEM)'. *Catena* 37(3–4), 431–447.

Tam, W.Y. 1997: 'Pattern formation in chemical systems: roles of open reactors'. In, H.F. Nijhout, L. Nadel and D.S. Stein (eds.), *Pattern Formation in the Physical and Biological Sciences,* Addison-Wesley, Reading, MA. 323–347.

Thomas, R. and Nicholas, A.P. 2002: 'Simulation of braided river flow using a new cellular routing scheme'. *Geomorphology* 43(3–4), 179–195

Thorn, C.E. and Welford, M.R. 1994a: 'The equilibrium concept in geomorphology'. *Annals of the Association of American Geographers* 84(4), 666–696.

Thorn, C.E. and Welford, M.R. 1994b: 'No dirge, no philosophy, just practicality'. *Annals of the Association of American Geographers* 84(4), 706–709.

Tucker, G.E. and Slingerland, R. 1996: 'Predicting sediment flux from fold and thrust belts'. *Basin Research* 8, 329–349.

Tucker, G.E. and Slingerland, R. 1997: 'Drainage basin responses to climate change'. *Water Resources Research* 33(8), 2031–2047.

Turkington, A.V. and Phillips, J.D. 2002: 'Cavernous weathering, instability and self-organisation'. Paper presented at RGS-IBG Session, *Geomorphology: Chaos, Fractals and Self-organizing Systems.* Belfast, January 2002.

Von Bertalanffy, L. 1968: *General Systems Theory: Foundations, Development, Applications,* George Braziller, New York.

Waldrop, M.M. 1994: *Complexity.* Penguin, London.

Washington, R.W. 2000: 'Quantifying chaos in the atmosphere'. *Progress in Physical Geography* 24(2), 499–514.

Watts, D.J. 1999: *Small Worlds: The Dynamics of Networks between Order and Randomness.* Princeton Studies in Complexity, Princeton, USA.

Werner, B.T. 1995: 'Eolian dunes: Computer simulations and attractor interpretation'. *Geology* 23, 1107–1110.

Werner, B.T. 1999: 'Complexity in natural landform patterns'. *Science* 284, 102–104.

Werner, B.T. and Fink, T.M. 1993: 'Beach cusps as self-organized patterns'. *Science* 260, 968–971.

Werner, B.T. and Hallet, B. 1993: 'Numerical simulation of self-organized stone stripes'. *Nature* 361, 142–145.

Wigner, E.P. 1960: 'The unreasonable effectiveness of mathematics in the natural sciences. *Communications in Pure and Applied Mathematics* 13(1).

Wilcox, B.P., Seyfried, M.S. and Matison, T.H. 1991: 'Searching for chaotic dynamics in snowmelt runoff'. *Water Resources Research* 27(6), 1005–1010.

Williams, J.R., Renard, K.E. and Dyke, P.T. 1983: 'EPIC – a new method for assessing erosion's effect on soil productivity'. *Journal of Soil and Water Conservation* 38(5), 381–383.

Wischmeier, W.H. and Smith, D.D. 1978: *Predicting Rainfall Erosion Losses.* US Department of Agriculture, Agricultural Research Service Handbook 537.

Wolfram, S. 1982: *Cellular Automata as Simple Self-Organising Systems.* Caltech preprint CALT-68-938.

Wolfram, S. 1986: 'Cellular automaton fluids 1. Basic theory'. *Journal of Statistical Physics* 45, 471–526.

Wooton, J.T. 2001: 'Local interactions predict large-scale pattern in empirically derived cellular automata'. *Nature* 413, 841–844

Wright, R. 2000: *Nonzero: The Logic of Human Destiny.* Little, Brown, London.

8
Cultural climatology

John Thornes and Glen McGregor

8.1 Introduction

Imagine that we know in advance that the earth's atmosphere is about to be poisoned by a giant meteor that will contaminate the air that we breathe. Attempts to destroy it have failed and we realize that until the atmosphere is made safe, and for the foreseeable future, we will have to breathe manufactured air. Animals and vegetation will also be affected, although it is not certain how. Huge factories to generate bottled air will have to be built and we will have to buy, for the first time, the air that we breathe. Currently, bottled air for scuba diving costs about £1/hour. Assuming that global capitalism could meet the world's needs at such a low price, how much would it cost to replace the atmosphere for a year?

Assuming the figures in Table 8.1 are in the right ballpark, consider the following. The global value of ecosystem services (ignoring the value of the air that we breathe) has been estimated to be of the order of US$16–54 trillion and, furthermore, the global GNP in 1994 was US$18 trillion (Costanza et al., 1997). Clearly then, the value of the atmosphere is far greater than the combined value of the global ecosystem and GNP. Indeed, it is much more valuable than we think. Not only does the atmosphere provide the air necessary to sustain life, but it is also the source of the supply of energy and moisture for food production and drinking water, a dumping area for pollutants and a medium for air transport for global travel and tourism. The atmosphere is, therefore, our most fundamental and valuable resource, which far outweighs our more normal perception of the atmosphere as a hazard that brings gales, floods, droughts, heatwaves

Table 8.1 *The value of the air that we breathe*

1. Cost of bottled air £1/hour = £ 8,760 /year

2. World population approximately 6.5 billion

3. Total cost 6.5 * 109 * 8,760 = 57 * 1012 = £57 trillion pounds = US$80 trillion dollars

4. The average person breathes up to 15 litres of air every minute which is nearly 8 million litres (approximately 10,000 kg) a year.

5. The atmosphere weighs approximately $5.3 * 10^{18}$ kg

6. The population of the earth breathes $6.5 * 10^9 * 10^4 = 6.5 * 10^{13}$ kg of air in one year

and freezing temperatures. Simply, without the atmosphere, life on earth would be impossible! The true economic value of the atmosphere is therefore way beyond the US$80 trillion calculated above. Yet we clearly take the atmosphere for granted. Perhaps we should all pay an 'air tax' to remind us of the atmosphere's indispensable nature and to help protect it and ensure its sustainability!

As geographers, it is surprising that we never study the atmosphere in this sort of relevant way. Rarely are questions asked such as: who owns the atmosphere? Is the atmosphere a common? How much is the atmosphere worth? How is the atmosphere used as a resource? How can we use our understanding of the atmosphere to alleviate atmospheric hazards and how can we ensure a sustainable atmosphere? These questions are of vital interest, not only to geographers, but also to world governments, global commerce and industry, and to each individual who faces such hazards as flooding, drought and storm. The recent problems with the Kyoto Agreement, which attempts to make a start on limiting enhanced global warming, highlights our ignorance and the need for further study of these cultural links.

Traditionally, geographers have studied the atmosphere through climatology as a branch of physical geography. This became known as geographical climatology, practised by geographer-climatologists (Carleton, 1999). This is to distinguish it from meteorological climatology, which is purely interested in the science of the atmosphere and rarely in the human impacts on or from the atmosphere. As the emphasis in climatology over the past 20–25 years has been traditionally biased towards meteorological climatology, the purpose of this chapter is to make a case for a more society or culture orientated climatology that considers the complex interactions between the climate system and the human social/cultural system. Before making this case, we will provide some scene-setting in terms of the meaning of climate and climatology and the relationship of climatology to the philosophy of science. We will do this in order to establish the main research themes geographer climatologists pursue and how they pursue them in terms of approaches to scientific explanation. This will be followed by a consideration of the abstract relationships between climate and society/culture. The chapter will close by assessing whether a cultural turn is required of climatologists as climatology becomes more concerned with the application of its science to the solution of climate-related environmental and societal problems and, therefore, needs to consider earnestly the relationship between climate and society.

8.2 Meanings of climatology in the last 25 years

The study of the atmosphere has never been more newsworthy and popular than in recent years. The study of climate, particularly has come to the attention of everyone for a number of reasons. Carleton states that the science of climate in the last 25 years has 'gained enormously in both scope and prestige, in stark contrast to its former role as the "ugly duckling"… when it was concerned largely with the compilation of station annual precipitation and temperature statistics' (Carleton, 1999, 713)

The traditional subgroups, methods and scales of climatology are shown in Figure 8.1, taken from Oliver (1981). These subgroups are hardly exciting and led to the description of climatology as a 'dry as dust' subject. In the last 25 years, this situation has changed considerably due, primarily, to the spur of climate change, but also other factors.

Carleton (1999) gives six main reasons as to why climatology has become so popular. These are:

- the recognition of an integrated climate system encompassing the biosphere, the cryosphere and the hydrosphere (especially the oceans)
- a perceived increase in both severe weather (tropical cyclones, tornadoes, heatwaves) and climate anomalies (e.g. droughts, extended wet periods)
- a growing awareness of the impact of human activities on the climate – especially enhanced global warming
- development of General Circulation Models (GCMs) of the climate based on the physical and mathematical underpinnings of meteorology and atmospheric science
- availability of new sources of data for studying the climate system e.g. satellites
- the growing availability of data on the internet and its ease of analysis using statistical packages on personal computers.

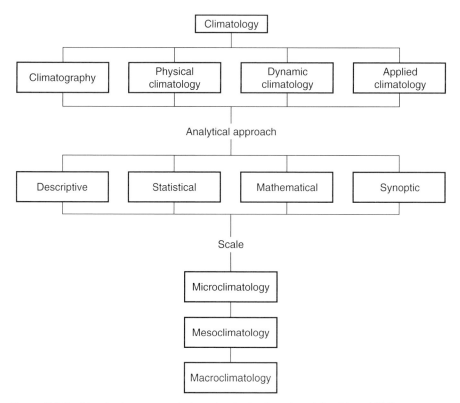

Figure 8.1 *Traditional subgroups, methods and scales in climatology (after Oliver, 1981).*

We may also add the following:

- the growing possibility of climate prediction using our knowledge of such events as El Niño (ENSO) and the North Atlantic Oscillation (NAO)
- publicity surrounding the need to understand the 'hole in the ozone layer'
- the recognition by industry and commerce of the importance of climate to their 'bottom line', for example, the establishment of a weather/climate 'futures' market and the importance of reinsurance to protect insurance companies from climate disasters
- publicity surrounding the so-called Kyoto Agreement to limit human-induced climate change.

The public awareness of climate change has increased exponentially in this period and yet the causes, impacts and management of climate change remain unresolved. They are still open for considerable further research. As O'Riorden (2000) states:

> *Climate change therefore will test science, science-policy relationships, global environmental agreements, the economics of response, the politics of coalition building across interests and generations, the morality of individual 'lifestyle choices' in the light of the innocence–blame contradiction, and the collective ethics of responding, coping, adapting and sharing in a world that is not yet one.*

Climatology has undergone a revolution – it is now global and cultural in scope. Have geographers been part of this revolution? Most certainly they have, but in doing so, have geograhers retained their geographical approach to the study of climate and climate processes? Carleton suggests that it is increasingly difficult to distinguish between the geographical and meteorological climatology research undertaken today, due to expanded interdisciplinary and multidisciplinary cross-fertilization of ideas. To lend a distinct geographical bent to climatology, Carleton (1999, 721) has proposed that geographers should take a lead and use their skills to explore in more depth 'land surface–climate interactions', which are obviously spatial and can help to solve 'important questions of public concern with policy implications'. While such a call is applaudible, it ignores humans as occupiers of the land surface and as possible players in land surface climate interactions.

The definition of climatology has also changed significantly over the last 25 years. Bryson (1997) criticizes definitions such as 'climate is the synthesis of the weather', which is an abstract concept that implies that climate only exists as a statistical entity. Alternatively, he chooses to define climate with a small 'c' as ' the statistical assemblage of the weather in a region or at a place' (Bryson, 1997, 451). This is similar to the definition of climatography as 'the basic presentation of climatic data' (Figure 8.1).

Bryson goes on to suggest a new definition (which he calls Axiom 1) of Climate with a big 'C': 'Climate is the thermodynamic/hydrodynamic status of the global boundary conditions that determine the concurrent array of weather patterns'.

This reverses the abstract notion that climate is 'average weather' and suggests that weather is restrained by climate within a 'season's allowable array'. He goes on to define three more axioms and six corollary statements, as shown in Table 8.2.

This definition of climate in 'axiom 1' treats climatology as the study of the impact of the surface of the earth (boundary conditions) on the atmosphere. This is similar to Carleton's 'land surface–climate interactions'. Interestingly, like Carleton (1999), Bryson excludes any mention of human culture in his definition of climatology. This also appears to be the case with a recent re-definition of synoptic climatology offered by Yarnal et al. (2001) such that synoptic climatology 'integrates the simultaneous atmospheric dynamics and coupled response of the surface environment'. Although justified by its 'focus on application, rather than methodology' (Yarnal et al., 2001) this definition excludes or, at best, only implies that society and culture may be part of the surface environment.

In short, climate–society relationships and the implications these may hold for the management of climate-related environmental and socio-economic problems is not an integral concept in current definitions of climatology. Perhaps there is a need to

Table 8.2 *The Axioms and Associated Corollaries* of Climatology (Bryson, 1997).*

Axiom 1	Climate is the thermodynamic/hydrodynamic status of the global boundary conditions that determine the concurrent array of weather patterns.
Corollary 1.1	The initial emphasis of climate theory and associated modelling should be on the conditions of the most important boundaries.
Corollary 1.2	One need not model the weather day by day for many years equivalent and then sum to model the climate.
Corollary 1.3	Climate is multidimensional (a vector), not a single scalar datum.
Axiom 2	The history of climate is a non-stationary time series.
Corollary 2.1	There are no true climatic 'normals'.
Axiom 3	Environment and climate change on timescales from near instantaneous to millions of years.
Corollary 3.1	There can be no perfect climatic or environmental analogs in the last million years. Conversely, reconstruction of past climates must be based on methods that do not require perfect analogs.
Axiom 4	There are various microclimates that depart more or less from the macroclimates in each terrestrial region.
Corollary 4.1	It follows that proxy data that are very local, that is, dependent on the local micro-climate, may not reflect the true macroclimate.

*An axiom is a statement that is accepted as true without proof or argument.
A corollary is something that is evident after something has been proved.

re-define climatology with a focus on the physical (atmosphere, hydrosphere, cryosphere, biosphere, land surface) and human (cultural, social, political, economic) components of the climate system such that climatology is:

> *the study of the processes of, and the interactions and feedbacks between, the physical and human components of the climate system at a variety of temporal and spatial scales.*

This definition acknowledges foremost that humans are an integral part of the climate system, but in doing so, emphasizes that through cultural, social, political and economic activities, society responds to and/or influences the physical nature of the climate system through interactions and feedbacks. Not only is the nature of the processes, interactions and feedbacks of concern to the geographical climatologist, but also the outcomes of these, as manifested by the mean, variability and extreme characteristics of a location's or region's climate and the resultant ways in which society utilizes, exploits and adapts to these characteristics. Furthermore, the outcomes of the processes, interactions and feedbacks may vary over a range of frequencies (slow to fast variations), throughout human history, and from place to place, such that climatology, as a discipline, recognizes that the two-way relationship between society and the physical components of the climate system is non-stationary through time and over space.

In short, we advocate that climatology should not only be concerned with the study of physical processes at various space and time scales, but with evaluating and understanding climate society interactions and feedbacks as made manifest by societal response and how society may interpret climate information. Therefore, we view climate as an integral part of culture and, as such, we contend that there is a need to develop a sub-discipline within climatology that we will refer to as cultural climatology.

In order to understand how climatology functions as a science and to provide a philosophical platform for justifying a call for cultural climatology, the philosophy of climatology and the nature of the relationships between climate and society will be discussed first.

8.3 Climatology and the philosophy of science

A philosophical disposition that dominated the basic sciences throughout the latter and first halves of the 19th and 20th centuries respectively was that of positivism. Pursued by the logical positivists, this placed emphasis on the importance of positive facts and observable phenomena and became the building blocks of what is widely referred to as scientific method. The positivist stance was characterized by empirical generalizations and law-like statements that relate to empirically recognizable phenomena. As noted by Gregory (2000), positivism embodies the verification principle because it relies on empirical generalizations to propose testable empirical hypotheses that may be verified or falsified. The aim of such hypothesis testing, and thus the positivist approach, was the generation of general laws that are not circumstance specific. This philosophy pervaded the sciences so that by the early 1950s, the function of science was seen as to: 'establish general laws covering the behaviour of the empirical events or objects with which the science in question is concerned, and thereby to enable us to connect together our knowledge of seperately known events, and to make reliable predictions of events as yet unknown' (Braithwaite, 1953, 1).

Such a *raison d'être* certainly applied to climatology and was at the heart of much climatological research, as it is today. Although the basic purpose of science was not in contention, the procedures used by the logical positivists in producing the law-like statements were subject to criticism. The main arguments against the logical positivist approach were that experience, the use of the senses and inductive argument were applied in the construction of scientific knowledge. Furthermore, the use of repeated observations to produce general statements was flawed because observations are theory laden and sound principles of verification and induction were lacking. An outcome of these criticisms was the development of critical rationalism. This placed emphasis on the notion that a theory is assumed to be true until it is falsified. In contrast to the logical positivists, falsification was at the heart of theory testing, not verification. In the words of Gregory (2000, 50):

> the scientist proposes trial solutions which are then evaluated critically, the trial solutions being speculative theories set up to solve the particular problem to hand. Falsification is justified as a procedure because, whereas no finite number of facts can verify a universal proposition, a single fact can demonstrate the proposition to be false. Scientific statements are therefore conceived as being falsifiable whereas non-scientific ones are not.

In climatology and the atmospheric sciences as a whole, the trial solutions mentioned above are basically models that embody theories and provide a basis for making predictions about the outcome of measurements (Randall and Wielicki, 1997). In this sense, a model is a story about the atmosphere and that story must be consistent with the measurements of the atmospheric phenomena of interest. A model can also be considered as a hypothesis because a model can be used to perform a calculation that produces a prediction about the state of the atmosphere. Therefore, phrased in the form of a hypothesis, it could be said that the model predictions are thought to be true. This statement can then be falsified by comparing the model output with actual atmospheric measurements. If there is conflict between the model and the measurements, then the model can be considered as falsified (Randall and Wielicki, 1997).

Models, which may take on one of four forms, namely elementary models, forecast models, models that simulate statistics directly and toy models (Randall and Wielicki, 1997), have become of prime importance in the atmospheric sciences. As noted by Robinson and Henderson-Sellers (1999, 2) 'a major aim of contemporary climatology is to predict future climatic conditions'. Such predictions are usually based on climate modelling, which still relies largely on a positivist doctrine.

From a philosophical standpoint, Peterson (2000) argues that the use of climate models and computer simulations to produce future climatic scenarios may be challenged for several reasons. His concerns relate to the:

- spatial and temporal scales needed to accurately simulate global climate change at a 100-year timescale

- assumptions and simplifications (parameterizations) used in all climate models that render their results speculative

- the tuning of climate models which artificially prevents a model from obtaining an incorrect result and thus limits research progress which should be targeted at falsifying models and understanding the exact nature of the processes being modelled

- verification as opposed to falsification of models as verification of a model is logically impossible (Popper, 1959)

- a lack of acknowledgement of model uncertainty, which should be of great concern to policy makers. Usually modellers stress the virtues of their models and not the uncertainties. Policy makers and the media want single number answers and it is tempting to forget the error bars (Rind, 1999).

Despite the above problems, most of the climatological community is striving to produce models, whether they be conceptual (Nicholson & Gris, 2001), statistical (Zweirs and von Storch, 1999) or numerical (McGuffie and Henderson-Sellers, 2001), in order to describe and explain the behaviour and causal mechanisms of some climate or climate-related phenomena.

In relation to geomorphology, Richards et al. (1997) have proffered that explanation frequently involves application of methodologies that shift between the dichotomies of positivism and realism. This is also apparent in climatology. Positivitism connotes an empirical and positivist approach based upon observational and experimental evidence that relies on sampling theory, large sample sizes, statistical methods and empirical generalization. Such an approach is dominant in climatology and manifests itself in statements such as: 'Almost all climatologists, whatever their training, employ some variant of the composite or ensemble average to assist in generalization as opposed to the single case study', and 'In climatological enquiry undertaken by geographers, ... composites provide a basis for interpreting detailed case studies, rather than letting the case studies inform the general model' (Carleton, 1999, 714).

Robinson and Henderson-Sellers (1999), in discussing the six idealized stages in the development and occlusion of a depression, also note that the average picture is of greater emphasis in climatology compared to meteorology: '... as climatologists rather than meteorologists, we are mainly concerned with identifying general characteristics of depressions, not the particular characteristics of any single one' (Robinson and Henderson-Sellers, 1999, 156). Similarly, this positivist approach, based on large sample sizes and generalization, is embodied in the concept of the canonical or typical or average El Niño event (Glantz, 2001). However, as noted by Allan (2000) in relation to the average climatic patterns and impacts associated with El Niño: 'although composites of climatic patterns and impacts during El Nino and La Nina events tend to be of the opposite sign to one another, individual El Nino or La Nina events are never the same and can vary in magnitude, spatial extent, onset, duration, cessation, etc' (Allan, 2000, 4).

The fact that there may be variants on the composite or average model, plus there is a very low probability of actually observing the average pattern or number (Zweirs and von Storch, 1999), implies that for many climatic phenomena, no one dominant climatic pattern may exist. For this reason, many climatologists, still within the context of positivitism, have turned to the use of statistical data reduction techniques, such as principal components analysis, which rely on large sample sizes, to isolate the dominant modes of climatic variability at a variety of spatial and temporal scales.

A further characteristic of the positivist philosophy that is predominant in climatology is the application of the deductive approach to scientific explanation. This contrasts with the inductive approach or received view stance (Gregory, 2000) adopted by the logical positivists, and is more akin to the model theoretic view (Rhoads and Thorn, 1996). The deductive approach is based on testing an a priori model of a pattern, process or relationship in the real world or a theoretical statement that defines a family of theoretical models that represent real-world phenomena. Arising from the a priori model may be a number of working hypotheses that are assessed in terms of their falsifiability by applying statistical analyses to the comparison of real-world experimental/measured data with the precepts embodied in the theoretical statement or a priori model. Non-falsification of the hypothesis leads to acceptance of the working hypothesis, lending credence to the a priori model or theory. As noted by Yarnal (1993), synoptic climatologists follow a deductive approach to scientific explanation as hypotheses concerning atmospheric circulation and environment linkages are erected and tested by comparing the variability of a surface environmental parameter, such as air quality (McGregor and Bamzelis, 1995) or human mortality (McGregor, 1999), with a descriptor of the state of the atmospheric circulation. The deductive approach is also visible in other fields, such as urban climatology, in which it is commonly hypothesized that the urban heat island form will match closely the physical urban form, or that latent heat flux, as an energy sink, is a fundamental control on urban/non-urban climate differences. Dynamic climatology (often referred to more recently as climate dynamics), the basis for early atmospheric general circulation models (Rayner, 2000), on the other hand, applies an inductive approach in attempting to mathematically describe the characteristics of the general atmospheric circulation from first principles (Yarnal, 1993).

In contrast to positivism, which appears to dominate climatology, is realism and the realist approach to scientific explanation (Richards et al., 1997). This places emphasis on the case study or a small number of cases, the objective being to develop an explanation of the mechanisms that generate the phenomenon under investigation. Richards et al. (1997) note that three levels can be distinguished for a phenomenon. These are the underlying mechanisms and the intellectual structures representing them; events caused by those mechanisms in particular circumstances; and observations of those events (Gregory, 2000). Based on the above it would appear that the realist approach is more typical of meteorology, in which special field observation programmes are planned to uncover the exact mechanisms and processes underlying a particular atmospheric phenomena. The recent FASTEX experiment (Joly et al., 1999), in which an intensive programme of observation and subsequent modelling was undertaken to increase the meteorological community's understanding of the genesis of North Atlantic storms, is a good example of this approach. Although predominant in meteorology, the realist approach is not totally absent in climatology as intensive process-based case studies can help inform the climatologist about the processes that give rise to general climatic patterns. For example, much of the early case study-based work on urban energy balances (Oke, 1982 and Arnfield, 2002), which helped explain observed temperature distributions in cities, were in this vein.

A theme that has permeated discussions on the philosophy of science in general is that of the paradigm. Attributable to Kuhn (1962), a paradigm is viewed as a widely recognized scientific achievement that for a period provides model problems and solutions to a community of practitioners. Further, Kuhn proposed a conceptual model of how science develops in a series of phases. A pre-paradigm phase would precede a stage of professionalization, when the definition of the subject is well honed. Subsequently, a series of paradigm phases would follow, characterized by scientists seeking solutions to problems within generally accepted rules and conventions. Such activity was referred to as 'normal' science. However, as problems arose with the prevailing paradigm, a crisis point would be reached and revolution would follow. As a consequence, the scientific discipline would, in effect, be re-set to a new pre-paradigm phase and development would proceed in a phased manner again until the next 'crisis'. Although the general applicability of the paradigm model in physical geography has been questioned (Stoddart, 1986), within climatology, a number of persistant paradigms can be identified.

Perhaps the most sustained paradigm is that of the Bergen school's cyclonic model, which not only has been the basis of short-term weather forecasting and analysis for over 75 years, but also has been applied to the interpretation of variations in surface environmental parameters (Comrie, 1990 and Yarnal et al., 2001). Moreover, it forms the conceptual basis of both manual and computer-assisted classifications of the atmospheric circulation (Yarnal, 1993). Other paradigms that dominate climatology presently are the ocean–atmosphere paradigm and the paradigm of the climate system. The former has provided a context within which climatologists have made rapid progress in understanding the nature and origins of the El Niño–Southern Oscillation (ENSO) phenomenon and, subsequently, the development of seasonal to inter-annual climate prediction models (Goddard et al., 2001).

A potted history of the scientific activity associated with El Niño (Glantz, 2001) is a good way of seeing how a paradigm develops, in this case, the ocean–atmosphere paradigm. The period prior to 1957 was characterized by a slow but, at times, faltering accumulation of scientific knowledge about El Niño. Despite Berlage (1927) implicating sea-surface temperature variations as a possible driver of atmospheric pressure variation across the Pacific (the Southern Oscillation, SO), this precept was not taken up as a theory until the late 1950s. In 1957, Berlage produced a comprehensive description of the Southern Oscillation and linked it to El Niño (Berlage, 1957). The 15-year period that followed witnessed an ever-increasing exposure of the El Niño phenomenon to the scientific community and the public. However, it was not until 1966 that an important milestone in El Niño research was achieved. Bjerknes (1966, 1969) explained the linkages between the SO and eastern equatorial Pacific sea-surface temperatures and demonstrated that El Niño was not confined to the coastal waters off Peru, but was a true basin-wide phenomenon and, more importantly, that the climate patterns associated with its occurrence were a result of interaction between processes in the ocean and the atmosphere. This gave rise to the now commonly used acronym of ENSO (El Niño Southern Oscillation), which has become embedded in the language of scientists, journalists and, to a large extent, the public. Bjerknes' 'discovery' catapulted the scientific community into a period of relentless research activity that continues to the present time (Diaz et al., 2001 and Goddard et al., 2001). Directed at developing a general ENSO theory, this activity spans the positivist and realist approaches to that development. In Kuhnian terms, ENSO science has almost, if not already, reached the stage of normal science, as the ENSO delayed oscillator theory has gained widespread acceptance amongst ENSO scientists (van Oldenburgh et al., 1999). ENSO research activity within the ocean–atmosphere paradigm over the last 30 years has also spawned interest in the analysis of the mechanisms underlying ocean–atmosphere interactions and their associated climatic impacts elsewhere, especially in the North Atlantic region (Marshall et al., 2001).

The ocean–atmosphere paradigm, and the disposition it represented in terms of explaining the workings of the ENSO phenomenon, was very much in line with ideas of systems theory, which gained ascendancy in the earth and environmental sciences throughout the 1960s and 1970s. Thus the ocean and atmosphere were viewed as open systems linked by exchanges of energy (sensible and latent heat), mass (water by precipitation and evaporation) and momentum (frictional drag). Systems thinking also set the context for the development of the climate system paradigm within which the climate modelling community currently works. However, it was not until almost 20 years after the climate system was formally defined by the World Meteorological Organization as composed of the atmosphere, hydrosphere, cryosphere, land surface and the biosphere (WMO, 1975), that the coupled or interlinked nature of the climate system was acknowledged in a formal definition. This definition, as set out in the 1992 United Nations Framework Convention of Climate Change, saw the climate system as 'the totality of the atmosphere, hydrosphere, biosphere and geosphere and their interactions' (McGuffie and Henderson-Sellers, 2001, 1068).

Whereas pure ENSO science is close to the Kuhnian idea of 'normal' science, this appears not to be the situation for climate-change science (Funtowicz and Ravetz, 1990). As outlined above, normal science is associated with one hypothesis or theory that strongly influences a discipline; uncertainties in the field of research are low. Furthermore, the knowledge generated within a discipline is adequate for addressing science and policy issues. In contrast, post-normal science is typified by the situation where normal science has generated a substantial body of knowledge in various disciplines, but there is a high level of uncertainty and great potential for disagreement due to empirical problems and political pressure (Bray and von Storch, 1999). Such attributes of uncertainty and high decision stakes appear to characterize current climate-change science according to the results of a survey of 400 climate scientists in Germany and North America:

> ...climate science has provided enough knowledge so that the initiation of abatement measures is warranted. However, consensus also exists regarding the current inability to explicitly specify detrimental impacts that might arise from climate change. This incompatibility between the state of knowledge and the calls for actions suggests that, to some degree at least, science advice is a product of both scientific knowledge and normative judgment, suggesting a socioscientific construction of the climate change issue.
>
> Bray and von Storch, 1999, 439

What is clear from this is that if climate-change science is to turn into effective climate-change policy, then climate-change scientists must transcend their traditional disciplinary boundaries and work closely with social scientists and policy makers. This would also appear to be the situation in the case of the climate-prediction community:

> If societal benefit is the goal of climate prediction, then it is clear that a strong orientation towards users, the decision-making process, and the social setting is required in applications research. There is a wealth of related information in the fields of development studies, global change, technology transfer and weather forecast application upon which to draw. Furthermore, experience with users forms a critical basis for guiding the priorities of prediction research...
>
> Goddard et al., 2001, 1140

Given that the ultimate goal of climatology is to apply the field's knowledge to the solution of both environmental and socio-economic problems, it appears that a cultural turn will be required of climatologists as they enter the post-normal phase of their science. The nature of this cultural turn, and what might constitute cultural climatology, will be discussed following a consideration of climate and society relationships as we consider these to be at the heart of cultural climatology. Furthermore, an evaluation of such relationships may assist with identifying the possible directions a cultural turn may take.

8.4 The relationship between climate and society

In order to provide a context for discussions relating to cultural climatology, it is perhaps important to first outline the general nature of the relationship between nature (climate) and society. This will aid in developing an understanding of how geographers currently study climate as a part of nature and where the opportunities may lie for developing cultural climatologies. The following draws heavily on Phillips and Mighall (2000), who have presented a discussion on the abstract relationships between nature and society. Here, their discussion is couched in terms of climate being a part of nature.

There are perhaps two possible ways in which nature may be viewed (Smith, 1984). One is that nature is external to human society and, thus, culture, such that climate, as part of nature, is seen as an object not influenced by human activity in any way. The other view contends that nature is an inherent state. Thus climate is universal and acts in a particular and unchanging way (Phillips and Mighall, 2000) and climate is ahistorical and does not change through history. Associated with the idea that climate as part of nature is an inherent state, is the further notion of one-dimensionality; that is, climate possesses a single basic character or 'essence' (Phillips and Mighall, 2000). Clearly, climate as a part of nature cannot be viewed dialectically as either external object or inherent state. This is because human activity can and has influenced climate either directly or indirectly throughout human history (von Storch and Stehr, 2000). From an evolutionary and ecological perspective, the concept of climate as external object can also be questioned as humans have evolved and developed as a part of nature (Phillips and Mighall, 2000). Furthermore, climate as an inherent state is doubtful, as the globe's climate has been characterized by major episodes of glaciation, interspersed with interglacials. Moreover, between the glacial–interglacial episodes have been periods with climates substantially warmer than the present interglacial. As noted by Harvey (2000, 1), the climate is a 'dynamically, constantly changing phenomenon'.

In addition to considering how nature or climate might be thought of, it is important to acknowledge that there are two contrasting philosophies concerning how society might relate to climate as part of nature. These are the philosophies of determinism and free will. Phillips and Mighall (2000) refer to these as society equals climate (deterministic) and climate equals society (free will). The society equals climate position signifies the idea that society is determined by climate. This position rejects the concept of climate as an external object and accepts the idea of climate as inherent state; that is, society and everything else is viewed as being determined by climate and its inherent characteristics. This contrasts with the free will or human agency stance that rejects both the ideas of climate as external object and inherent state. Therefore, in contesting the idea that climate equals society, the free will/human agency position implies that climate is seen to have no meaning; simply, it does not exist. Climate (nature) is a consequence of human agency and, in a sense, is replaced by society. As noted by Phillips and Mighall (2000), people who adopt this perspective are referred to as social constructionists. In this sense, climate is viewed as socially created.

To this point society has been referred to as meaning people in general. However, society can be thought of in two main ways: society as the outcome of the actions of individuals and society as a collective of notions or forces. These are referred to as the individualist and structuralist perspectives on society. They carry with them quite different interpretations of society–climate relations (Phillips and Mighall, 2000). An individualist perspective of society–climate relationships would place emphasis on personal beliefs about climate and the consequences of an individual's actions on climate. For example, to what extent might an individual think global but act local in terms of climate change. The answer to this will depend on personal beliefs about the prospects of climate change, whether an individual believes they have any power to act and if they possess any vested interests. In contrast, the structuralist perspective is concerned with the general societal conditions under which people act, such as the existence of a climate change tax related to individual fossil fuel related energy consumption. Of course, it is not unimaginable that society and individuals will act upon each other such that there will be consequences for the individual as well as for the society to which they belong. This is the dialectical conception of society that places emphasis on a two-way interaction between society and individuals (Phillips and Mighall, 2000).

If the contrasting viewpoints on nature/climate are integrated with the three conceptions of what constitutes society, then a framework for considering attitudes on climate, society and society–climate relations emerges. This is presented in Figure 8.2.

Clearly the society equals climate and climate equals society are diametrically opposed standpoints. Geographers in the study of climate and society rarely adopt these extreme approaches. As noted by Phillips and Mighall (2000), geographers usually occupy the middle ground between these dipoles. Such a position reflects debates of the late 19th and early to mid-20th centuries between environmental determinists and cultural determinists or possibilists. The former group, as exemplified by the German geographer, Ratzel and his American follower, Semple, adhered to the notion that man is a product of the earth's surface and, as reflected by the theories of Ellsworth Huntington, the course of human civilization is closely tied to climate and climate change (Johnston, 1987). The cultural determinists or possibilists countered these arguments by insisting that humans were, in fact, active as opposed to passive agents. They exercized their free will in choosing the most appropriate use of the environment that suited their

View of interrelation of society and climate	Society is part of nature/climate	Society and nature/climate are distinct but interlinked objects		Nature equals society/climate
	Holistic	Dualistic		Holistic
View of causal linkages	Climate creates society	Climate dominates society	Society dominates climate	Society creates climate
View of 'climate as external object'	Rejects	Accepts	Accepts	Rejects
View of 'climate as universal attribute' or inherent object	Accepts	Accepts	Rejects	Rejects
Philosophical viewpoint	Environmental constructionism	Environmental determinism	Social determinism	Social constructionism
View of society	View of climate/society			
Individualistic view	Individuals have behaviour determined by climate	Individuals have behaviours, determined by climate but these may be modified by an individual	Individual perceptions and behaviour impact behaviour as determined by climate	Individual perceptions and actions construct climate
Dialectical view	Individuals and societies adopt laws of climate through interaction with each other through processes of nature	Individuals/ societies have behaviours determined by climate, but these may be modified by social norms/ individual agency	Individual perceptions and actions within social structures impact the climatic environment	Individual perceptions and actions within social structures construct climate
Structuralist view	Societies adopt laws of climate	Societies are conditioned by climate and affected by social norms	Social structures impact the climatic environment	Social structures construct climate

Figure 8.2 *Attitudes on climate and society (after Phillips and Mighall, 2000).*

cultural outlook. In doing this, humans were seen as agents creating nature *not* being products of it.

From a climate point of view, the determinist stance is probably easier to defend in very general terms as there are ample examples of ways in which climate sets the broad environmental boundary conditions for a range of human activities and behaviours. For example, not withstanding psychological determinants, clothing design characteristics vary with extreme climatic conditions. Housing style also varies in a broad way with climate: compare Mediterranean and Northern European house styles (Mather, 1974). There is also clear evidence from the reinsurance industry that weather/climate high-risk areas incur greater insurance premiums compared to low-risk areas (Chagnon et al., 1997). In contrast, the contention that climate is socially constructed is less easy to support if we think of climate in pure physical terms as measured by the standard climate variables. However, one may argue that at the micro-level, air conditioning or heating is a form of social construction of climate in that indoor climate is artificially created but actively controlled by human intervention. Also, add to this the fact that socio-economic status may also influence climate at such a micro-level as the ability to create an equitable indoor climate will be intimately related to one's access to such control as determined by finances or the general social environment. Such access may also determine one's vulnerability to the health effects of extreme hot or cold events (Smoyer, 1998). Urban climates can also be considered as socially constructed, as urbanization, a social process, alters the thermodynamic, moisture and radioactive properties of the earth's surface and thus creates climates that contrast with adjacent non-urban areas. This is a direct parallel with the social construction of landscapes through, for example, agricultural practices. At the global scale, perhaps, it can also be contended that through increasing greenhouse gas concentrations, human society is constructing climate.

Social construction not only relates to physical alteration of the climate but also to the notions of climate possessed by individuals and societies. For example, the perception of climate may well be related to the extent to which a society or individual has experienced emancipation from the influences of climate. Urban dwellers spend much of their time indoors. For most of the time they are immune to the effects of climatic vagaries. What is more, their livelihoods are not likely to be tied directly to climatic variations, as would be the case for rural workers. Therefore, social context may play an important role in determining what an individual or society knows about climate. In effect, climate may be socially constructed, as climate information received by society is rarely unmediated. It is a filtered version of that produced by climate scientists. This, in turn, may affect the societal response to climate policy (Stehr and von Storch, 1995). Not only might the level of direct experience of climate determine climate knowledge, but also the scientific community is influential. This is especially so with regards to knowledge concerning future climate related to an enhanced greenhouse effect. If a constructivist approach to climate science is adopted (Jasanoff and Wynne, 1998), then climate knowledge is not only scientifically constructed but may also be moderated by a number of cultures, including bureaucratic, economic and civic cultures (Figure 8.3). Culture here means a combination of knowledge, ideas and beliefs, together with discourses (agreed interpretations of contentious issues), practices and goals. Hence science is modified by culture and is constantly being re-defined or reconstructed (hence the term constructivist approach). If we apply this constructivist model to climate change

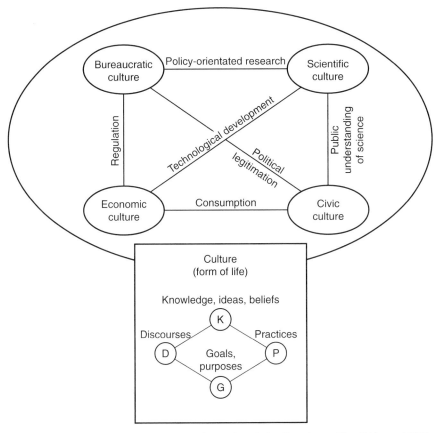

Figure 8.3 *Moderating cultures operating on climate knowledge (from Jasanoff and Wynne ,1998).*

then we can begin to understand how difficult it is to get global agreement on the Kyoto Protocol.

8.5 Climate and the cultural turn

Within the social sciences a distinct 'post-modern turn', and within human geography a manifest 'cultural turn', have been noted in the 1990s. Mitchell (2000, 59) is wary of this cultural wonder and comments:

> Indeed, some see the 'postmodern turn' in social sciences and humanities – with its multifaceted concern with (and some would say uncritical wonder at) all matters cultural, and its retreat from studies of economic systems and processes of exploitation, coming as it did just as the political and economic right gained ascendancy – as marking a rather complete surrender to the forces of reaction.

Culture is a complex and overused word, and is used to encapsulate everything from the 'total way of life' of a people (language, dress, religion, music, values etc.) through to works of art and advertising. Mitchell (2000, 14) gives six possible definitions of culture, but the simplest definition of all, which has been used here and is possibly of most use to climatologists, is: 'Culture is all, that is not nature'.

Although the definition of nature is not as simple as it might seem (Macnaghten and Urry, 1998), we can fruitfully examine this dialectic between climate as part of nature and culture. The critical examination of the impact of climate on culture (Thornes, 1999; Janković, 2000; and Hamblyn, 2001) and the impact of culture on climate, could be seen to be the new role of the geographical climatologist or cultural climatologist.

Returning to Figure 8.2, this presents a useful framework for considering the rubric of approaches the geographer might take in evaluating climate and society/culture relationships and thus developing cultural climatologies. The extreme bottom right-hand corner of the figure represents the interaction between the structuralist view of society and the philosophical outlook of social constructionism. This interaction provides a context within which geographers might explore to what extent social structures such as class, the media, education systems and their associated curricula and scientific institutions construct climate. For example, there is evidence that the way in which the media portrays climate has an impact on how society as a whole perceives the current climate state (Henderson-Sellers, 1998 and Ungar, 1999). Over-reporting on climate extremes, for example, can lead to a false impression of how variable the climate is and the sole attribution by society of climatic variability to an enhanced greenhouse effect. By placing recent media reports within a longer-term climatological context, geographers could make a useful contribution to deconstructing media-constructed images of climate. Social structures such as scientific institutions also construct climate through climate predictions. Useful comparisons could be made of not only how such climate forecasts are constructed, but also the underlying socio-economic and, perhaps more importantly, the political nuances that determine the nature of the forecasts. The same exercise could also be conducted with reference to climate change predictions, especially with regard to the assumptions underlying climate change scenarios and their faithfulness to reality. Investigating how the social structures of class, race, ethnicity and gender determine images of climate also represents a useful pursuit for geographer-climatologists. This is because perceptions of climate will influence how sectors of society may respond to climate-related policy and climate information 'handed down' from scientific institutions.

Continuing with the structuralist view of society, but in the context of a social determinist view of climate, the extent to which social structures, such as the International Monetary Fund and the World Bank, through providing finance for large development projects impact climate is an area awaiting investigation by climatologists. Studies of the way in which large dam, irrigation or agricultural development projects might influence local to regional scale hydroclimatological processes could be undertaken by running numerical regional climate models. In the context of climatic determinism and a structuralist view of society, are questions such as: how do social and cultural norms act as filters or modifiers of general climatic constraints on the behaviour of societies? With respect to the interaction of environmental constructionism and structuralist views of society, are questions such as: to what extent are various societies emancipated from climate? Or from an individualistic/environmental constructionalist viewpoint (top left hand corner), how climate or weather sensitive are individual members of society? Related to this is the issue of the extent to which individual weather/climate sensitivity may be modified by individual behaviour (environmental determinism/individualistic interaction). For example, does the extent to which someone is cold sensitive determine whether holidays are taken in warmer climes at the time of the year when coldest temperatures can be expected? How does an individual modify their behaviour to avoid weather and climate that is to their dislike? In the context of the dialectic view – environmental determinism interaction – how does an individual's circumstances within a given cultural setting or social norm enable them to modify or respond to a general sensitivity to weather and climate through the ability to install air conditioning, take a holiday in a much more equitable climate or modify their dress style? At the level of the individual, and taking a social constructionist perspective (top right-hand corner), are questions concerning climate and trust; that is, does society trust the climate to behave in a normal fashion? For example, is society's trust in climate as a faithful agent (Stehr, 1997) compounded by the fact that following weather or climate extremes, climate and socio-economic conditions return to normal sooner or later? Moreover, if so, can an apparent trust in the normality of climate be broken if politicians, scientists and the media continue to give conflicting views or, in other words, construct images of climate that conflict with those held by individuals?

Although not exhaustive, the above are some of the issues that geographer-climatologists could well begin to address within the context of cultural climatology. Just as climatology, in the 'normal science' sense, is concerned with micro to global scale climate, cultural climatology should be concerned with studying climate society relationships from the macro scale down to that of the individual and asking questions about how these may vary through time (changing experiences) and between cultures and social structures. However, to pursue cultural climatology, a new wave of climatologists equipped with social science analysis skills will be required. This has been recognized for some time. In 1976, the geographer and climatologist, Werner Terjung (Terjung, 1976), proposed a 5-level systems approach (Figure 8.4) to show how geographical climatologists could use their social science skills to try and develop deductively constructed physical-human process-response systems (level 5). Although Terjung provided guidance on how this could be achieved: 'Prospective physical geographers should take basic courses in calculus, physics, chemistry, engineering, modern biology and computer programming. On higher levels of instruction geography departments should develop courses, which stress a core of basic thermodynamics and hydrodynamics and their relation to the environmental envelope of relevance to

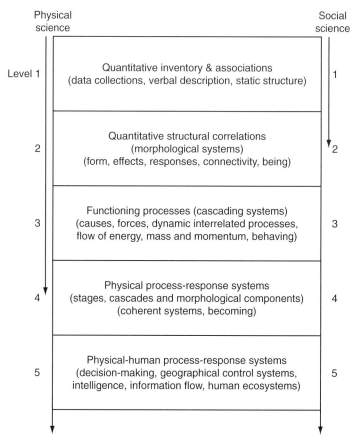

Figure 8.4 *Cultural climatology level 5 = physical science level 4 + social science level 2 (after Terjung, 1976).*

mankind', (Terjung, 1976, 221), no hint was given concerning how climatologists would acquire the necessary 'human' social science skills. In describing the make-up of level 5, he states: 'The flow of information in decision-making systems would be analysed on the level of morphological systems (level 2) which are then linked to environmental process-response systems' (level 4). Thus the geographical climatologist is expected to achieve level 2 in social science and level 4 in physical science in order to be able to achieve solutions to real-world problems, which are at level 5.

Thornes (1978, 1981) compared this level 4 + 2 approach to the concept of atmospheric management, which combines an understanding of atmospheric science and perception of the atmosphere in order to solve applied climatological/ meteorological problems (Figure 8.5). Today we might state that the modern cultural climatologist attempting to tackle level 5 problems needs to be trained in physical science to level 4 and social science to level 2. Cultural climatology level 5 = physical science level 4 + social science level 2.

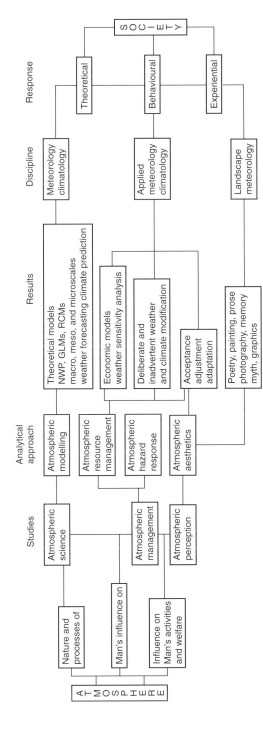

Figure 8.5 A simple model of approaches to atmospheric studies (after Thornes, 1978).

In a similar vein, but much more recently, Perry (1995, 281), has noted: 'if the "complete climatologist" of the future is an individual who not only has had a thorough grounding in atmospheric science studies but is also conversant in the social sciences, this has major implications for the training and perhaps the background of the new generation of climatologists'.

Perhaps the new generation of climatologists referred to by Perry (1995) could well be called cultural climatologists. The chief interests of such a new generation should be understanding and explaining climate and society relationships within the rubric of 'society' and philosophical outlook interactions, as expressed in Figure 8.2. In undertaking cultural climatological studies, it is likely that a variety of paths will be followed in pursuing explanation of the linkages and feedbacks between climate and society. We imagine a dualistic approach in which methodologies will migrate between positivism and realism with, perhaps, deductive approaches dominating. Moreover, the outcomes of cultural climatological studies are expected to be models or 'stories' that offer insight into climate and society interactions. Such models, whether they are conceptual, statistical or even numerical, will form the building blocks of decision support systems that will allow the testing of decision options for societies at large or individuals.

8.6 Cultural climatology and climatology

We would like to emphasize that the summons to cultural climatology, presented here, should not be misconstrued as a call for climatologists to abandon the research mainstays of synoptic, dynamic and physical climatology. On the contrary, we expect and encourage geographer-climatologists to continue to be concerned with issues such as global climate system change, establishing what the principal drivers of the climate system are, assessing how the climate system will respond to natural and human-induced changes, answering how society might respond to the opportunities and threats posed by climate change, and evaluating to what extent the changes expected in the climate system can be predicted. This call to cultural climatology should be seen more as a signal to climatologists that opportunities await us at the interface between science and society, an area which physical geographers on the whole have felt great apprehension with. In the words of Terjung (1976, 222):

> Because of a predilection for self-contemplation and sometimes almost suicidal academic isolation, geography has a history of missing the boat on vital social and environmental issues. Many of these problems could be handled, ideally, by geographers who have a broad grasp of the world around us. We have long held out the promise of the integrative examination of man (society) and nature (climate) – here is our opportunity (bracketed words are ours)

We see cultural climatology as just the ticket for catching the climate and society boat of opportunity. A passage on this boat will bring us closer to understanding the physical and societal mechanisms underlying the complex interactions between components of the climate–human system. We therefore invite all students and purveyors of climatology courses to look beyond the learning and teaching of straight climate processes by considering the multitude of ways in which climate and society may interact. Such a broadening of our horizons into the realms of cultural climatology will not only provide us with learning and research opportunities, but will also provide society with a better understanding of the meaning of climate.

References

Allan R.J. 2000: 'ENSO and climatic variability in the last 150 years'. In Diaz, H.F. and V. Markgraf (eds.), *El Nino and the Southern Oscillation: Multiscale Variability and Global and Regional Impacts*. Cambridge University Press, Cambridge.

Arnfield A.J. 2002: 'Two decades of urban climate research: small scale climate processes and the urban heat island'. *International Journal of Climatology* 22 (in press).

Berlage, H.P. 1927: 'East Monsoon forecasting in Java'. *Verhandelingren Koninklijk Magnetisch en Meteorologisch Observatorium te Batavia*, 20.

Berlage, H.P. 1957: 'Fluctuations of the general atmospheric circulation of more than one year, their nature, and prognostic value'. Royal *Netherlands Meteorological Institute Yearbook*, 69, 151–159.

Bjerknes, J. 1966: 'A possible response of the atmospheric Hadley circulation to equatorial anomalies and ocean temperature'. *Tellus*, 8, 820–829.

Bjerknes, J. 1969: 'Atmospheric teleconnections from the equatorial Pacific'. *Monthly Weather Review*, 97, 163–172.

Braithwaite, R.B. 1953: *Scientific Explanation*. Cambridge University Press, Cambridge.

Bray D. and von Storch, H. 1999: 'Climate science: an empirical example of post-normal science'. *Bulletin of the American Meteorological Society*, 80, 439–455.

Bryson, R.A. 1997: 'The paradigm of climatology: an essay'. B*ulletin of the American Meteorological Society*, 78, 449–455.

Carleton, A.M. 1999: 'Methodology in climatology'. *Annals of the Association of American Geographers*, 89, 713–735.

Chagnon, S.A., Chagnon, D., Fosse, E., Hoganson, D.C., Roth Sr, R.J. and Totsch, J.M. 1997: 'Effects of recent weather extremes on the insurance industry: major implications for the atmospheric sciences'. *Bulletin of the American Meteorological Society*, 78, 425–435.

Comrie, A.C. 1990: 'The climatology of surface ozone in rural areas: a conceptual model'. *Progress in Physical Geography*, 14, 295–316.

Costanza, R., d'Arge, R., de Groot, R., Farber, S., Grasso, M., Hannon, B., Naeem, S., Limburg, K., Paruelo, J., O'Neill, R.V., Raskin, R., Sutton, P. and van den Belt, M. 1997: 'The value of the world's ecosystem services and natural capital'. *Nature* 387, 253–260.

Diaz, H.F., Hoerling, M.P. and Eischeid, J.K. 2001: 'ENSO Variability, Teleconnections, and Climate Change'. *International Journal of Climatology*, 21(15), 1845–1862.

Glantz, M.H. 2001: *Currents of Change: Impacts of El Nino and La Nina on Climate and Society*, Cambridge University Press, Cambridge.

Goddard I., Mason, S.J., Sebiak, S.E., Ropelewski, C.F., Basher, R. and Cane, M.A. 2001: 'Current approaches to seasonal-to-interannual climate predictions'. *International Journal of Climatology*, 21, 1111–1152.

Gregory, K.J. 2000: *The Changing Nature of Physical Geography*. Arnold, London.

Haines-Young, R. and Petch, J. 1986: *Physical Geography* – Its Nature and Methods. Harper & Row, Cambridge.

Hamblyn, R. 2001: *The Invention of Clouds*, Picador Press, London.

Harvey, L.D. 2000: *Global Warming: The Hard Science*. Prentice Hall, Harlow.

Henderson-Sellers, A. 1998: 'Climate whispers: media communication about climate change'. *Climatic Change*, 40, 421–456.

Janković, V. 2000: *Reading the Skies*. University of Manchester Press, Manchester.

Jasanoff, S. and Wynne, B. 1998: 'Science and decision making'. In Raynor, S. and Malone, E. (eds.), *Human Choice and Climate Change: the Societal Framework*. Battelle Press, Columbus, Ohio, 1–87.

Johnston, R. J. 1987: *Geography and Geographers : Anglo-American Human Geography Since 1945*. Edward Arnold, London.

Joly, A. et al. 1999: 'Overview of the field phase of the fronts and Atlantic storm-track experiment (FASTEX) project'. *Q.J.R. Meteorol. Soc.*, 125, 3131–3164.

Kuhn, T.S. 1962: *The Structure of Scientific Revolutions*. University of Chicago Press, Chicago.

Macnaghten, P. and Urry, J. 1998: *Contested Natures*. Sage, London.

Marshall, J., Kushnir, Y., Battisti, D., Chang, P., Czaja, A., Dickson, R., Hurrell, J., McCartney, M., Saravanan, R. and Visbeck, M. 2001: 'North Atlantic climate variability: phenomena impacts and mechanisms'. *International Journal of Climatology*, 21 (in press).

Mather, J.R. 1974: *Climatology : Fundamentals and Applications*. McGraw-Hill, New York.

McGuffie, K. and Henderson-Sellers, A. 2001: 'Forty years of numerical climate modelling'. *International Journal of Climatology*, 21, 1067–1110.

McGregor, G.R. 1999: 'Winter ischaemic heart disease deaths in Birmingham, UK: a synoptic climatological analysis'. *Climate Research*, 13, 17–31.

McGregor, G.R. and Bamzelis, D. 1995: 'Synoptic typing and its application to the investigation of weather air pollution relationships, Birmingham, UK'. *Theoretical and Applied Climatology*, 51, 223–236.

Mitchell, D. 2000: *Cultural Geography*, Blackwell, Oxford.

Nicholson, S.E. and Gris, J.F. 2002: 'Conceptual model for understanding rainfall variability in the West African Sahel on interannual and interdecadal timescales'. *International Journal of Climatology*, 21(14), 1733–1757.

Oke, T. 1982: 'The energetic basis of the urban heat island'. *Quarterly Journal of the Royal Meteorological Society*, 108, 1–24.

Oliver, J.E. 1981: *Climatology: Selected Applications*. Edward Arnold, London.

O'Riordan, T. 2000: *Environmental Science for Environmental Management*, Prentice Hall, Harlow.

Perry, A.W. 1995: 'New climatologists for a new climatology'. *Progress in Physical Geography*, 16, 97–100.

Peterson, A.C. 2000: 'Philosophy of climate science'. *Bulletin of the American Meteorological Society*, 81, 265–271.

Phillips, M. and Mighall, T. 2000: *Society and Exploitation Through Nature*. Prentice Hall, Harlow.

Popper, K.R. 1959: *The Logic of Scientific Discovery*. Hutchinson, London.

Randall, D.A. and Wielicki, B.A. 1997: 'Measurements, models and hypotheses in the atmospheric sciences'. *Bulletin of the American Meteorological Society*. 78, 399–406.

Ravetz, J. 1990: 'Knowledge in an uncertain world'. New Scientists, 127, 18.

Rayner, J.M. 2001: *Dynamic Climatology*. Blackwell, Massachussets.

Rhoads, B.L. and Thorn, C.E. 1996: *The Scientific Nature of Geomorphology*. John Wiley, Chichester.

Richards, K.S., Brooks, S., Cliford, N., Hams, T. and Lane, S. 1997: 'Theory, measurement and testing in "real" geomorphology and physical geography'. In D.R. Stoddart (ed.), *Process and Form in Geomorphology*. Routledge, London.

Rind, D. 1999: 'Complexity and climate'. *Science*, 284, 105–107.

Robinson, P.J. and Henderson-Sellers, A. 1999: *Contemporary Climatology*. Longman, Harlow.

Smith, A.G. 1984: Uneven Development: Nature, Capital and the Production of Space. Basil Blackwell, Oxford.

Smoyer, K.E. 1998: 'A comparative analysis of heat waves and associated mortality in St. Louis, Missouri – 1980 and 1995'. *International Journal of Biometeorology*, 42(1), 44–54.

Stehr, N. 1997: 'Trust and climate'. *Climate Research*, 8, 163–169.

Stehr, N. and von Storch, H. 1995: 'The social construct of climate and climate change'. *Climate Research* 5, 99–105.

Stoddart, D.R. 1986: *On Geography*. Basil Blackwell, Oxford.

Terjung, W.H. 1976: 'Climatology for geographers'. *Annals of the Association of American Geographers*, 66, 199–222.

Thornes, J.E. 1978: 'Applied climatology: atmospheric management in Britain'. *Progress in Physical Geography*, 2, 481–493.

Thornes, J.E. 1981: 'A paradigmatic shift in atmospheric studies?' *Progress in Physical Geography*, 5, 429–440.

Thornes, J.E. 1999: *John Constable's Skies: A Fusion of Art and Science*, University of Birmingham Press, Birmingham.

Ungar, S. 1999: 'HYPERLINK "http://kapis1.wkap.nl/oasis.htm/184995" Is strange weather in the air? A study of US national network news coverage of extreme weather events'. *Climatic Change*, 41, 133–150.

van Oldenborgh, G. Jan., Burgers, G., Venzke, S., Eckert, C. and Giering, R. 1999: 'Tracking down the ENSO delayed oscillator with an adjoint OGCM'. *Monthly Weather Review*, 127, 1477–1496.

Von Storch, H. and Stehr, N. 2000: 'Climate change in perspective'. *Nature*, 405, 615.

WMO 1975: *The Physcial Basis of Climate and Climate Modelling*. GARP Publication Series No. 16. WMO/ICSU, Geneva.

Yarnal, B. 1993: *Synoptic Climatology in Environmental Analysis: A Primer*. Belhaven Press, London.

Yarnal, B., Comrie, A.C., Frakes, B. and Brown, D.P. 2001: 'Developments and prospects in synoptic climatology'. *International. Journal of Climatology* (in press).

Zwiers, F.W. and von Storch, H. 1999: *Statistical Analysis in Climate Research*. Cambridge University Press, Cambridge.

9

Geomorphological and landscape wisdom – using local knowledge to manage slopes

Gerardo Bocco and Juan Pulido

9.1 Introduction

Traditional knowledge has been defined as the local knowledge of a cultural group or a society, as opposed to the global, universal knowledge generated by universities, research institutes and other formal institutions (Warren, 1991), usually through the testing of scientific hypotheses (Zimmerer, 1994). Although different terminology may suggest conceptual differences, in this chapter, the terms traditional, local, rural, peasant and indigenous are used interchangeably to describe local, non-scientific knowledge.

Traditional knowledge of the environment is holistic, integrating many scientific fields, including climatology, hydrology, soil science and plant and animal ecology (Howes and Chambers, 1980; Jungerius, 1985; Wilken, 1987 and Agrawal, 1995). Several researchers have highlighted the usefulness of traditional knowledge in rural sustainable development for developing countries, because of its inherent ecological foundation (Jungerius, 1985; Bocco, 1991; Altieri, 1992; Lamers and Feil, 1995 and Toledo, 1998). For example, traditional agriculture uses local farming practices that have been developed over thousands of years (Pulido and Bocco, 2003).

One subject that is seldom referred to in the literature is that of traditional, peasant geomorphological/landscape knowledge. Most of the research has concentrated on biological or agronomical topics. This chapter looks at the contribution of geomorphological wisdom to landscape and slope management in order to define universal principles. It specifically looks at:

- the major contributions to local soil and land knowledge, as described in the literature

- peasant geomorphological/landscape and local slope knowledge

- two examples from Mexican indigenous communities in the central part of the country

- the ongoing use of local knowledge in indigenous land-use planning for Oaxaca, Mexico.

9.2 Local soil knowledge

Traditional soil knowledge has been widely documented (Williams and Ortiz-Solorio, 1981; Bocco, 1991; Pawluk et al., 1992; Zimmerer, 1994; Lamers and Feil, 1995; Sandor and Furbee, 1996; Krogh and Paarup-Lawrsen, 1997 and WinklerPrins, 1999). WinklerPrins (1999) defines local soil knowledge as the knowledge of soil properties and management possessed by people living in a particular environment for some period of time.

Local classification systems commonly exhibit non-exclusive taxonomic relations, a characteristic that differs from the hierarchical structure of scientific taxonomies; this suggests that local understanding of the biophysical world often differs from the conventional ones of science and scientists (Zimmerer, 1994). However, the structure and refinement of local soil classification systems have been discussed thoroughly, and their use has been suggested for improving technical soil surveying in terms of accuracy, cost reduction and permitting rapid participatory land evaluation procedures (Howes and Chambers, 1980; Williams and Ortiz-Solorio, 1981; Millington, 1984; Sandor and Furbee, 1996; Ortiz-Solorio, 1999 and Veihe, 2000).

Conversely, the use of modern technology to fully understand and validate traditional knowledge has also been suggested for land management purposes (Bocco and Toledo, 1997 and WinklerPrins, 1999). However, controversy has arisen because some researchers question the idea that scientific knowledge can be superior to its traditional counterpart (see the discussion in WinklerPrins, 1999).

The need to study and document traditional rural knowledge in different ecological and cultural environments has also been emphasized in the literature (Krogh and Paarup-Lawrsen, 1997; Duffield et al., 1998; Gliessman, 1998 and Dickson, 1999). But, as stated by WinklerPrins (1999), there has been little published that is comprehensive or unified, probably because of the variety of subjects involved and the scattering of the papers in different scientific publications and 'grey' literature. One important effort is the work of Barrera and Zinck (2000), who made a thorough review of ethnological experiences throughout the world.

Many investigators have discussed conceptual and practical relationships between traditional rural knowledge and modern technologies, such as geographical information systems (GIS), database management and expert systems (Agrawal, 1995; Zwahlen, 1996; Bocco and Toledo, 1997; Brodnig and Mayer, 2000 and Gobin et al., 2000). In this approach, local knowledge is accepted as a robust source of data to support decision-making.

As pointed out by some researchers (Agrawal, 1995 and Zwahlen, 1996), local knowledge is not a panacea for every problem in sustainable rural development. However, its potential contribution to the management of current natural resources, especially in developing countries, has been agreed in the literature (Dumanski, 1997; Duffield et al., 1998 and Brodnig and Mayer, 2000).

9.3 Peasant geomorphological/landscape knowledge in Mexico

Territorial segmentation by rural producers is commonly hierarchical. The largest geographic and productive area recognized by peasants is their own community or *ejido.* In Mexico, ejidos are the land-tenancy areas recognized for rural producers (*mestizos* or indigenous) by the Constitution. These areas are unique and labelled with a specific geographical name. If relief amplitude permits, the ejidos are divided by peasants into bioclimatic categories: *tierra caliente, tierra templada* or *tierra fría* (literally, warm, temperate and cold lands). In tropical and subtropical regions, there is a dramatic change in environmental conditions following significant changes in altitude within short horizontal distances. These areas demonstrate contrasts in natural vegetation, soils and farming systems. It is quite common that, within a single ejido, in hilly areas in central and southern Mexico, relief amplitude may cover a couple of thousand meters. Therefore, in a hypothetical case, wood and resin may be extracted from temperate forests at higher altitudes; settlements maybe located in the middle fringe, where subsistence agriculture of maize and beans takes place; and coffee, banana and other tropical fruits may be cropped at lower altitudes (below 1200 m above sea level).

Every tierra is divided into localities (*parajes*). Parajes are unique and are also labelled with a specific geographical name. Their boundaries are well defined in terms of agricultural management, such as cattle grazing, forestry or protected area (for a spring, for instance). In the case of milpa production, a paraje is composed of several small areas of land (parcels) that are managed similarly in terms of their cropping calendar and cattle grazing (Pulido and Bocco, 2003).

Localities are, in turn, divided into peasant landscape units (PLUs). These are areas of land that the peasant recognizes and differentiates from other areas of land because of location and attributes such as relief, soil quality, hydro-meteorological conditions, landuse and management (Pulido and Bocco, 2003). PLUs correspond to the land units defined by landscape ecologists (*sensu* Zonneveld, 1995).

PLUs are divided into slope sequences following relief characteristics. An evaluation of soil and soil quality is attached to every slope topological sequence. Major peasant decision-making processes are based on land information at this scale. Three types of land quality are vaguely recognized by peasants (Openshaw and Openshaw, 1997): good, medium and poor, varying according to organic matter content, slope gradient and slope position. Good land is darker whilst poor land is lighter peasants directly relate colour to soil fertility. This reasoning matches the classification of land quality characteristics in formal land evaluation procedures (Rossiter, 1990 and Gobin et al., 2000).

Landforms and land-cover units are also described using peasant-defined categories. Slopes are described as *filos* or *filetes* (sharp terrain), whereas plains are called *parejos* or *planes* (levelled terrain) and depressions. The terms *mesa*, *cerro* and *loma* (the former for a small plateau and the second and third modest elevations) are both local and scientific labels. *Monte* (mount) is used to describe an elevation with a forest cover; *jaral* describes a shrubland within agricultural lands, following long-term fallow periods; and *pajonal* is referred to as a grassland.

9.4 Traditional landscape knowledge in Nuevo San Juan, Mexico

Nuevo San Juan Parangaricutiro is an indigenous (Purepecha ethnic group) community, located 15km west of Uruapan, in the state of Michoacán (Figure 9.1). The area covers roughly 1000m of relief amplitude (between 1800 and 2850m above sea level). Therefore, climate is temperate, with a seasonal rainfall of about 1000mm (yearly average). The major economic activity of the community is temperate forest management. At least 60% of its nearly 18,800ha is covered by mixed forest of *Pinus sp.* and *Quercus sp.* Landforms and soils have recent volcanic origin; the most prominent feature is the Paricutin volcano, a monogenetic cone that last erupted in 1943.

In the southwest, rain-fed agriculture (milpa, with maize and, occasionally, beans) is practised using traditional techniques. Pulido and Bocco (2003) describes thoroughly the local farming system in Nuevo San Juan; here, the landscape component of the system is reported. Soil was sampled (surface and subsurface layers) along slope transects, in three different peasant-defined quality classes, in 10 different parajes, covering the internal variability of the study site. The field research was carried out entirely with local producers (at least half the total population of 30 producers). Questionnaires were used to survey local knowledge of landscape segmentation and quality (Pulido and Bocco, 2003).

Figure 9.2 describes an idealized landscape transect with scientific and local (including Purepecha) names for categories of landforms, land-cover and land-quality. Soil analyses (Tables 9.1 and 9.2) indicated that land quality correlated fairly well with slope position and different yield indicators. Therefore, the local farming system is based on a thorough consideration of slope characteristics and processes. In addition, soil and water conservation techniques, such as earth-dams and channels protected with local vegetation species of shrubs and grasses, involve slope considerations.

Figure 9.1 *Location map of study sites.*

Local name	joya	cerro	ladera	mesa	joya	cerro	mesa	cerro
Purhépecha designation	Juata terhutzicurinï		Andáchukunï		Terhúnguinï		Tamban	Juata turhípitï
		Juata		Tamban	Juata sapich'u			
Paraje	"Cerro la chimenea"		"El tepamal"		"San Nicolás"		"Llaguacuaro"	"Cerro prieto"
Typical peasant land quality	Good/medium		Poor/medium		Poor		Poor	Medium/good
Soil unit (FAO)	Vitric andosol/Vitric eutric andosol		Eutric haplic andosol/Vitric eutric andosol		Eutric haplic andosol		Eutric haplic andosol	Vitric eutric andosol/Vitric andosol
Typical soil cover (technical)	Forest, maize		Maize, shrubs		Mize, shrubs and forest		Shrubs, forest	forest
(peasant)	Monte, maíz		Maíz, jaral		Maíz, jaral and monte		Jaral and monte	monte

Figure 9.2 *Idealized landscape transect in Nuevo San Juan with scientific and local (including Purhépecha) names for landforms, land-cover and land-quality categories.*

Table 9.1 *Yield and productivity of maize per land quality class (LQS).*

LQS	Yield of maize (Grain yield) kg ha⁻¹	Net aerial primary productivity (NAPP) kg ha⁻¹	Grain/NAPP Ratio %
Poor	1,290 (752)	3,163 (1,106)	37.5 (13.6)
Medium	2,090 (589)	4,707 (742)	44.6 (9.7)
Good	2,375 (941)	5,882 (1,597)	39.5 (13.89)
AVERAGE	1,918 (876)	4,584 (1,616)	40.54 (12.38)
MSD	816	1,570	9.51

Note: Data is the average of 8 estimations of NAPP following procedures by Ovington (1963) and Mueller-Dombois and Ellenberg (1974). Figures between brackets indicate standard deviations; *MSD*, minimum significant difference (source: Pulido and Bocco, 2003).

9.5 Indigenous slope knowledge in central Mexico

Many studies have reviewed indigenous slope knowledge and management over time. Field (1966) reported on terrace agriculture in Atlixco, Mexico. Herold (1966) and Dennis and Griffin (1971) discussed the use of *trincheras* as either terraces or check-dams for erosion control in the Mexican western Sierra Madre. Flannery et al. (1967) and Spores (1969) analysed local conservation techniques used by the Mixteco Indians in the Nochixtlan Valley, Mexico. Donkin (1979) studied agricultural terracing in tropical hilly countries. García (1986) discussed the development of pre-Hispanic conservation techniques in, respectively, Tlaxcala, Mexico.

Table 9.2 *Most significant correlation coefficients for some chemical and physical parameters of topsoil (0–25cm) and maize yield components. (α = 0.001 or better). (Pearson Correlation Coefficients / Prob > |R| under Ho: Rho=0 / N = 24).*

	SP	pH	Ca	OM	P	NAPP	PD
P	0.7162	–	–	–	–	–	–
Ca	–	0.8235	–	–	–	–	–
Mg	–	0.7137	0.8130	–	–	–	–
CEC	–	–	0.7006	0.7272	–	–	–
SLT	0.6416	–0.6237	–	–	–	–	–
NAPP	0.7018–	–	–	–	0.7592	–	–
GY	–	–	–	–	0.6014	0.8459	–
GY/NAPP	–	–	–	–	–	–	–0.7184

Note: SP, slope position; Ca, exchangeable calcium; OM, organic matter content; P, phosphorus; NAPP, net aerial primary productivity; PD, plant density; Mg, exchangeable magnesium; CEC, cation exchange capacity; SLT, 'sandy' layer thickness; GY, grain yield (taken from Pulido and Bocco, 2003).

Tapaxco is an ejido of indigenous (Mazahua group) influence. It is located in the Mexican volcanic belt, a Quaternary volcanic area, where at least half of the Mexican population resides (Figure 9.1). Climate is temperate, sub-humid with contrasting dry winters and rainy summers. Major landforms in the study area are gently sloping pyroclastic hills; the main soil type is planosols; and land use is rain-fed agriculture and grazing. Bocco (1991) described traditional techniques for soil and water conservation that were successful enough to recover an area severely gullied. This chapter reports on the local knowledge contained in those techniques and field surveys applied within the gully, carried out in eroding areas surrounding the gully system, and in still non-gullied terrain.

The results obtained from that survey were (after Bocco, 1990):

1. Techniques inside the gullies:

   (a) Disposal of crop residues at gully heads to decrease runoff speed in the approach channel and to collect sediment.

   (b) Construction of layered, earth/shrub (and stone, if available) filtrating check-dams to control channel activity and to trap sediment, especially in tributaries of the main channels. Locally, the dams are called *retranques* and are similar to the trincheras of indigenous groups in the southwestern United States/northwestern Mexico.

   (c) Planting long-rooted grass (>20cm, locally called *pasto carpeta*, or carpet grass) on gully walls to improve soil strength and to protect the walls from rainfall impact. Peasants recognize the risk of grass spreading to nearby areas of land as a major problem to further ploughing.

   (d) Breaking seepage scarps to decrease their height and angle and to increase stability.

2. Techniques on eroding slopes:

   (a) Disposal of scattered residue and stones on bare soils or exhumed subsoils to decrease the speed of water flowing overland (overland flow) and trap sediments coming from upslope. After some soil material is captured, rows of nopal (*Opuntia sp.*) or maguey (*Agave sp.*) are planted to create colluvial steps. The material is usually brought by hand or on animals from minute dams or holes dug to trap sediments. A 25cm thick layer of trapped soil was observed after three years of treatment.

   (b) Ploughing the sod on undulating, grassed areas, especially where tributaries start, to stimulate the removal of the surface horizon upslope by overland flow and to collect it further downslope. Collection is done by piling the sod obtained upslope and creating retranques downslope. After several years, the original irregular slope is levelled by erosion/deposition and can be ploughed.

   (c) Constructing small ditches to divert overland flow and collect sub-surface flow. The *charcos* (elliptical depressions, seasonally flooded) formed along the seepage scarps are thus drained.

3. Preventive techniques on (as yet) unaffected land:

   (a) Ploughing parallel to contours to varying depths, according to the irregularities of the slope, to control water movement. The peasants describe this activity as controlling the 'gravity' of the milpa.

   (b) Disposal of excess water from the milpa by constructing a number of ditches (10–15 m apart, parallel to contours) to avoid drainage concentrating in single spots. The main ditches are protected with *nopales* or *magueyes*.

   (c) Separation of milpas from each other by grassed areas 1–2m wide, to avoid the concentration of drainage and to create relatively protected footpaths.

Other techniques could probably be added to this list. The ones summarized here, however, were sufficient to reclaim severely eroded terrain. The techniques indicate a thorough understanding of slope dynamics, including (at least intuitively) sub-surface hydrology. This knowledge is part of the local culture, passed from one generation to the next (the area has a strong Mazahua influence). In other places in central Mexico, this kind of knowledge has been systematized and formalized by the indigenous groups. The Aztecs and the Purhépechas, for example, have developed their own soil taxonomies (Williams and Ortiz-Solorio, and Barrera, 1988).

The main principle behind most of the knowledge described in the literature seems to be the management of sedimentation (rather than erosion). In some instances, erosion is even stimulated upslope to capture sediments downslope. Such a technique was described for the Mixteca Alta (AD 1000) Oaxaca, Mexico (Kirkby, 1972 and Whyte, 1977), and suggested in studies of pre-Christian agricultural terraces in the Negev and

Judaen hills (Donkin, 1979, 33), as well as in local conservation practices in Sierra Leone (Millington, 1984) and Ghana (Veihe, 2000). Managing sedimentation means active and efficient slope management. Intuitively, at least, an important part of geomorphological knowledge on slope form and slope processes is embodied in this approach. In other cases, slope management is devoted to valley bottoms.

Discussions with farmers indicated that they perceived soil detachment as a process difficult or impossible to deal with and directly related to the effect of rainfall/runoff. Precipitation cannot possibly be controlled. Climate, especially rainfall, is of paramount importance in the daily life of agrarian societies. It plays a major role in ancient beliefs that still survive and are often masked by Christian practices (Bocco, 1990).

9.6 Traditional landscape knowledge for peasant land-use planning

Land evaluation for rural land-use planning is becoming a key tool for sustainable development in Third World countries. In Cochabamba, Bolivia, for instance, traditional knowledge is being used by governmental organizations to assess the quality of land in villages and for planning further development and food security (Mandujano, 2000).

In Mexico, major efforts have been made during the last decade to develop and reinforce land-use planning. However, data on land capabilities is not available at village level for the entire country. Since 75% of the forest resource (an umbrella-resource for soils and water) is accessed by indigenous forest communities and ejidos, the potential for peasant-defined, land-quality units is of paramount importance.

In Sierra Juarez, Oaxaca (Figure 9.1), several indigenous (Zapoteco group) communities have been successful in managing forests and agriculture without compromising their resource base. In some of these communities, land-use planning is being developed on the basis of traditional landscape knowledge (Ramírez, 2001). Peasant-defined, land-quality units are field surveyed, and databases are structured around combining potential versus actual land use. By doing this, conflicts can be detected, desired uses can be proposed and, in participatory workshops, either accepted or rejected by the community members.

Ongoing projects include training in cartography, aerial photography and GIS for resource management among technicians and high-school students of Sierra Juarez forest communities (Bocco et al., 2001).

9.7 Conclusions

This chapter discusses the concept of traditional geomorphological knowledge for landscape and slope management. Traditional knowledge is of a local nature. However, research suggests that some universal principles may be acting and guiding local knowledge in different parts of the developing world and over different periods of time. One such principle is of prime geomorphological importance and deals with the action of gravity forces along slopes. In every reported slope conservation technique, in different ecological environments, the control of sedimentation seems to be crucial. In addition, the segmentation of the territory in a hierarchical manner and the definition of slope facets and their labelling in terms of their productive quality also seem to be based on geomorphological reasoning.

9.8 Summary

Traditional knowledge has been defined as the local knowledge of a cultural group or a society as opposed to the global, universal knowledge generated by universities, research institutes and other formal institutions, usually through the testing of scientific hypotheses. In this chapter, the contribution of geomorphological wisdom to traditional landscape and slope management has been discussed in order to define the potential universal principles that underpin local knowledge. First, major contributions in local soil and land knowledge, as described in the literature, are analysed. Second, peasant geomorphological/landscape and slope knowledge are discussed, and two examples from Mexican indigenous communities in the central part of the country are reviewed. Finally, ongoing use of the approach in indigenous land-use planning is briefly described for Oaxaca, Mexico.

Acknowledgements

This chapter is based on research funded by projects PAPIIT IN-101196 and IN-101900. Heberto Ferreira gave technical support during several steps of the research. The second author was granted an MSc scholarship from CONACYT, Mexico, during 1997–2000.

References

Agrawal, A. 1995: 'Indigenous and scientific knowledge: some critical comments'. *Indigenous Knowledge and Development Monitor,* 3, 3–5.

Altieri, M.A. 1992: 'Sustainable agriculture development in Latin America: exploring the possibilities'. *Agriculture, Ecosystems and Environment,* 39, 1–21.

Barrera, N. 1988; 'Etnoedafología purhépecha'. *México Indígena,* 24, 47–52.

Barrera, N. and Zinck, J. A. 2000: *Ethnopedology in a Worldwide Perspective: an Annotated Bibliography.* International Institute for Aerospace Survey and Earth Sciences (ITC), Enschede.

Bocco, G. 1990: 'Gully erosion analysis using remote sensing and geographical information systems'. Unpublished PhD, University of Amsterdam, Amsterdam.

Bocco, G. 1991: 'Traditional knowledge for soil conservation in central Mexico'. *Journal of Soil and Water Conservation* 46(5), 346–348.

Bocco, G., Rosete, F., Bettinger P. and Velazquez. A. 2001: 'GIS program development with community participation in a developing country'. *Journal of Forestry* (in press).

Bocco G. and Toledo, V. M. 1997: 'Integrating peasant knowledge and geographic information systems: a spatial approach to sustainable agriculture'. *Indigenous Knowledge and Development Monitor* 5(2), 10–13.

Brodnig, G. and Mayer, V. 2000: 'Bridging the gap: the role of spatial information technologies in the integration of traditional environmental knowledge and western science'. *The Electronic Journal on Information Systems in Developing Countries* (available at http://www.unimas.my/fit/roger/EJ/SDC/vol1/vlrl.pdf).

Dennis, H.W. and Griffin. E.C. 1971: 'Some effects of trincheras on small river basin hydrology'. *Journal of Soil and Water Conservation*, 26(6), 240–242.

Dickson, D. 1999: 'ICSU seeks to classify "traditional knowledge"'. *Nature* 401, 631.

Donkin, R.A. 1979: 'Agricultural terracing in the aboriginal New World'. *Viking Fund Publications in Anthropology*, 56, The University of Arizona Press, Arizona.

Duffield, C., Gardner, J.S., Berkes F. and Singh, R.B. 1998: 'Local knowledge in the assessment of resource sustainability: case studies in Himachal Pradesh, India, and British Columbia, Canada'. *Mountain Research and Development* 18(1), 35–49.

Dumanski, J. 1997: 'Criteria and indicators for land quality and sustainable land management'. *ITC Journal*, 3–4, 216–22.

Field, C. 1966: 'Terrace agriculture'. Proceedings of the Latinoamerican Regional Conference. International Geographical Union, 2, 343–349. SMGE. Mexico.

Flannery, K.V., Kirkby, A.V., Kirkby M.J. and Williams. A.W. 1967: 'Farming systems and political growth in ancient Oaxaca'. *Science*, 158(3800), 445–454.

García, A. 1986: 'Control de la erosión en Tlaxcala: época pre-hispánica', (Erosion Control in Pre-Hispanic Tlaxcala). *Antropología*. INAH, Nueva Epoca, 10, 14–20. Mexico.

Gliessman, S.R. 1998: 'Agroecology: ecological processes in sustainable agriculture'. Sleeping Bear Press, California.

Gobin, A., Campling, P., Deckers J. and Feyen J. 2000: 'Integrated toposequence analyses to combine local and scientific knowledge systems'. *Geoderma* 97, 103–123.

Herold, L. 1966: 'Erosion control in Western Sierra Madre, Mexico'. Proceedings of the Regional Latinoamerican Conference. International Geographical Union, 2, 351–349. SMGE. Mexico.

Howes, M. and Chambers, R. 1980: 'Indigenous knowledge: analysis, implications and issues'. In D. Brokensha (Ed.), *Indigenous Knowledge and Development*. University Press of America, Lanham, MD. 323–3349

Jungerius, P.D. 1985: 'Perception and use of the physical environment in peasant societies'. *Geographical Papers No. 93*. Department of Geography, University of Reading, England.

Kirkby, M.J. 1972: 'The physical environment of the Nochixtlan valley, Oaxaca'. *Vanderbilt University Publications in Anthropology*, 2. Nashville.

Krogh, L. and Paarup-Lawrsen, B. 1997: 'Indigenous soil knowledge among the Fulani of northern Burkina Faso: linking soil science and anthropology in analysis of natural resource management'. *GeoJournal*, 43, 189–197.

Lamers, J.P.A. and Feil, P.R. 1995: 'Farmers' knowledge and management of spatial soil and crop growth variability in Niger, West Africa'. *Netherlands Journal of Agricultural Science*, 43, 375–389.

Mandujano, C. 2000: Senior scientist, UN Program on Food Security, La Paz, Bolivia. Personal communication.

Millington, A.C. 1984: 'Indigenous soil conservation studies in Sierra Leone'. *Bulletin of the International Association of Hydrological Sciences*, (IAHS). 144, 529–538.

Ortiz-Solorio, C.A. 1999: 'Los levantamientos etnoedafológicos'. (Ethno-edaphological Surveying). Unpublished PhD thesis. Colegio de Postgraduados en Ciencias Agrícolas, Montecillo, Mexico.

Openshaw, S., Openshaw. C. 1997: *Artificial Intelligence in Geography*. John Wiley, Chichester.

Pawluk, J., Sandor J. and Tabor J. 1992: 'The role of indigenous soil knowledge in agricultural development'. *Journal of Soil and Water Conservation*, 47(4), 298–302.

Pulido, J. and Bocco. G. 2003: 'The traditional farming system of a Mexican indigenous community: the case of Nuevo San Juan Parangaricutiro, Michoacan, Mexico'. *Geoderma*, III, 249–265.

Ramírez, G. 2001: 'Peasant-defined land-use planning in San Pablo Guelatao, Oaxaca'. *Proyectos de Desarrollo de la Sierra Norte, A.C.* Manuscript.

Rossiter, D. 1990: 'ALES: a framework for land evaluation using a microcomputer'. *Soil Use and Management*. 6(1), 7–20.

Sandor, J. and Furbee, L. 1996: 'Indigenous knowledge and classification of soils in the Andes of Southern Peru'. *Soil Science Society of America Journal*, 60, 1502–1512.

Spores, R. 1969: 'Settlement, farming technology and environment in the Nochixtlan Vally'. *Science*, 166 (3905), 557–569.

Toledo, V. 1998: 'Sustainable development at the village community level: a Third World Perspective'. In F. Smith (ed.), *Environmental Sustainability. Practical Global Implications*. St. Lucie Press, Boca Raton, 233–250.

Veihe, A. 2000: 'Sustainable farming practices: Ghanian farmers' perception of erosion and their use in conservation measures'. *Environmental Management*, 25(4), 393–402.

Warren, D.M. 1991: 'Using indigenous knowledge in agricultural development'. World Bank Discussion Paper 127, Washington, D.C.

Whyte, A.V.T. 1977: 'Guidelines for field studies in environmental perception'. UNESCO, MAB Technical Notes 5, Paris.

Wilken, G.C. 1987: *Good Farmers: Traditional Agricultural Resource Management in Mexico and Central America*. University of California Press, Berkeley and Los Angeles, CA.

Williams, B.J. 1975: 'Aztec soil science'. *Boletín del Instituto de Geografía*, 7, 115–120. UNAM. Mexico.

Williams, B.J. and Ortiz-Solorio C.A. 1981: 'Middle American folk soil taxonomy'. *Annals of the Association of American Geographers*, 71, 335–358.

WinklerPrins, A.M.G.A. 1999: 'Local soil knowledge: a tool for sustainable land management'. *Society & Natural Resources*, 12, 151–161.

Zimmerer, K. 1994: 'Local soil knowledge: answering basic questions in highland Bolivia'. *Journal of Soil and Water Conservation*, 49(1), 29–34

Zonneveld, I.S. 1995: *Land Ecology*. SPB Academic Publishing, Amsterdam.

Zwahlen, R. 1996: 'Traditional methods: a guarantee for sustainability?' *Indigenous Knowledge and Development Monitor*, 4(3), 18–20.

10
Conceptions of nature: implications for an integrated geography

Michael Urban and Bruce Rhoads

Introduction

Traditionally, the discipline of geography has laid claim to a holistic perspective on the world, while at the same time struggling with certain deep-seated internal dichotomies, none of which has been more troublesome than the split between human and physical geography. Despite an idealized notion that geography transcends the bounds of natural science, social science and the humanities, the discipline today functions more as two separate disciplines than as a coherent whole unified around a common core (Gober, 2000 and Johnston, 1997). A seemingly obvious common ground is the interface between humans and the biophysical environment, but few human and physical geographers currently are working together on problems centred in this interface (Rhoads, 1999). The claim that study of the interface between human society and environment lies 'at the very heart of geographical inquiry' essentially is a ruse; this interface has been and continues to be neglected by geographers, particularly physical geographers (Trimble, 1992).

For better or worse, physical geography is seen by physical and human geographers alike as being *different* from human geography. But what underlies this dichotomy? Is it simply a difference in subject matter, theoretical perspective, or research methodology? Are the two sides of the discipline looking for different types of knowledge? Do they have different conceptions of truth? Certainly the answers to these questions could all be yes, no or various mixtures thereof, depending on which geographers are providing the answers. Some may feel that contemplating such questions represents little more than an esoteric exercise in navel-gazing. What these geographers fail to realize is that refusal to engage the questions constitutes a response, one that others may feel justifies their perspective of geography as a superficial and perhaps superfluous academic discipline. The ways in which geographers conceptualize the scope of geographical inquiry 'conditions' and, in some cases, regiments practice (Sherman, 1996, 90). Such conceptions affect not only the process of obtaining, evaluating and accepting new geographical knowledge but, at a more fundamental level, mediate the questions we ask and the type of knowledge we seek (Bowler, 1993, 1 and Rhoads and Thorn, 1993, 290).

Philosophical examination of the relationship between physical and human geography has, until recently, not been of much interest to practising physical geographers (Chorley, 1978; Schumm, 1991; and Rhoads, 1999). For that matter, this issue has not been a focus of concern among human geographers either. Instead, this dichotomy has been ignored by both sides – human geographers, in many cases, writing treatises about 'geography' that either briefly mention or do not even acknowledge the existence of the physical geography and physical geographers, the minority, avoiding the discussion entirely. Meanwhile, the notion of an 'integrated' geography is promoted as propaganda within the discipline and serves as a façade to justify geography's distinctiveness to those outside the discipline.

This chapter examines philosophically the dichotomy between human and physical geography by arguing that it emanates from a traditional dualistic conception of the relationship between humans and nature. It shows how this dichotomy not only girds modern science, but also, in part, accounts for the division between the social and natural sciences, particularly in relation to investigations of environmental problems. The dualistic conception serves as an implicit constraint that scientists unwittingly impose upon themselves, thereby circumscribing notions of the 'acceptable' scope of scientific investigations in which they engage. Geographers, both human and physical, seem to subscribe to this presupposition, which accounts for the lack of integrative geographical investigations. A goal of the discussion is to illustrate for physical geographers that the emergence of a truly holistic geography requires both recognition of, and challenges to, the dualistic conception of humans and nature.

10.2 Contemporary physical geography and natural science

Contemporary research in physical geography can be characterized by a distinctive epistemological 'style' that determines appropriate ways of investigating and learning about the world (Richards, 1987 and Vicedo, 1995). This style cannot be encapsulated in a precise definition, nor does it encompass the activities of all physical geographers, but it does delineate the main focus of most studies. Since the 1950s, the standard of inquiry in physical geography has been a quantitative, process-oriented systems approach to the study of biogeographical, geomorphological and climatological systems (Gregory, 2000). Most investigations involve quantification, either in the development of theoretical constructs (analytical or numerical models) or in the analysis of empirical data (from field and laboratory studies). Investigations may follow deductive or abductive methodological frameworks (Rhoads and Thorn, 1993), but the emphasis is on understanding the processes or events that influence the development and evolution of earth's biotic, geomorphological and climatological systems (Rhoads and Thorn, 1996). To gain such understanding, physical geography draws upon background knowledge from a wide range of ancillary disciplines, including physics, chemistry and biology (Rhoads and Thorn, 1996 and Bauer, 1996). In this sense, it is similar in structure to related 'composite' sciences, such as ecology, that are composed of 'distinct parts of other types of study' (Osterkamp and Hupp 1996, 417). This strong dependency on theoretical knowledge from other natural sciences, many of which are viewed as foundational with respect to physical geography (Rhoads and Thorn, 1996), has led to an implicit acceptance by physical geographers of philosophical concepts underpinning natural science (Rhoads and Thorn, 1994) – a situation reinforced by casual assessments of physical geography by human geographers (Harvey, 1996, 60).

Environmental studies, or those that attempt to examine the role of human interaction with biophysical systems, also have become more prominent as a research theme over the past two decades. In physical geography, these studies usually adopt a perspective in which humans are separate from and external to 'natural' systems and thus represent a form of disturbance to these systems. A truly integrative geography should attempt to understand the interactions between society, the traditional focus of human geography, and the biophysical environment, the traditional focus of physical geography. Integrative research of this type largely represents a 'road not taken' by geographers (Trimble, 1992 and Rhoads, 1999). The lack of such investigations has its roots in assumptions about relations between humans and the biophysical world – relations that are subsumed by conceptions of nature.

10.3 Conceptions of nature

Identifying and exploring the conceptual boundary between humans and the biophysical environment is a complex exercise. This boundary typically has been explored within the context of the concept of nature, a term loaded with multifarious and sometimes contradictory symbolic content (Gifford, 1996, 33). While the existence of phenomena other than humans, referred to here as the biophysical environment, generally is not disputed, the meaning of the concept of 'nature' commonly is.

The relation of the concept of nature to human culture transcends simple taxonomy. Haraway (1994, 63) refers to the mythical, textual, technical, political, economic and organic dimensions of nature 'collapsing into each other in a knot of extraordinary density.' Colwell (1987) has argued that the debate about nature 'is more than the philosophers' contrivance; it is a fundamental expression of the entire sweep of modern culture.' Nature is one of the most complex words in the English language because it is linguistically unstable and has no absolute fixed referent (Williams, 1980 and McIsaac and Brun, 1999). Underlying the debate over this concept is a dispute about 'the legitimate and logically consistent ways to view human activities with respect to the rest of the natural world' (McIsaac and Brun, 1999, 2).

Nature can be seen as a group of interrelated objects that objectively exist in the world, independent from humans, along with the biophysical processes that create and maintain the objects and interrelations, or as a conception that is socially created and maintained by the human imagination (Soper, 1995, 136). However it is defined, nature is a powerful concept not only because it orders our perceptions of the world around us, but also because it delimits the relation between humans and non-human phenomena. Definitions of nature shape the spiritual and philosophical grounding of the human species. In the face of such hegemony, there is a temptation to simply eliminate usage of the term (Liu, 1989, 38 and McKibben, 1989). This avoidance, however, fails to confront the underlying relationship between humanity and the rest of the world. Nature as a concept 'needs to be contested, not rejected' (Bate, 1991, 56). Perhaps nowhere has the concept of nature been both embraced and contested more heartily than in the intellectual realm of Western culture. As stated by Lease (1995, 7), 'the place of humanity in nature – or, more precisely, the relationship of the human species to the rest of reality – has been a central problem in all historical cultures, but most particularly in the Western tradition that has produced both our contemporary "sciences" and the "humanities".'

10.4 The traditional conception: humans separated from nature

The legacy of nature that has emerged from Western thought is the result of thousands of years of debate and contestation (Fitzsimmons, 1989; Lease, 1995 and Castree and Braun, 1998). This legacy has been grounded largely in the philosophical stance that humans are separate from nature. According to this view, nature is the totality of biophysical processes and phenomena objectively unaffected by culture or the tainting influence of humans. A clear distinction is drawn between the natural, or 'other', and the human world.

The Western tradition of separating humans from the biophysical environment can be traced back at least as far as the 6th and 7th centuries BC (Glacken, 1967, 426). Since that time, people have detached themselves from the biophysical environment by invoking a number of essential dichotomies: *nómos* versus *physis,* body versus soul, and self-occurring phenomena versus the products of human artifice. The infusion of this conception into contemporary scientific perspectives emerged largely through the Christian view of the natural world and through Descartes' notion of mind–body dualism.

10.4.1 The Greek tradition

As early as the 5th century BC, Greek philosophy differentiated between what an object is and what it appears to be (Glacken, 1967, 51 and Soper, 1995, 37). The difference between the essence of phenomena and the way they appear or are perceived by humans is captured in the tension between *nómos* and *physis*. *Nómos* corresponds to sensory qualities we assign phenomena through vision, taste, smell and touch (Angeles, 1992, 205). *Physis* refers to primary qualities of phenomena such as size, shape and composition that exist independently of our interaction with the object (Angeles, 1992, 228 and Borgmann, 1995, 32). While *nómos* can vary from person to person depending on the way a phenomenon is perceived by an individual, *physis* embodies the true human-independent character of the phenomenon.

The original purpose of the *nómos/physis* distinction was to differentiate what is inherently real from what is perceptual or interpretive, not to distinguish people from an objective natural 'other.' However, *physis* gradually came to embody the ultimate reality or the true nature of things. The term often is used synonymously with nature or the natural order, effectively setting up a fundamental dichotomy between nature and humanity, even though the early Greeks considered *techné,* all the creations of humanity, to be a subset of *physis.* The pre-Socratic notion of the term 'nature' refers to the 'All or Everything' (Tuan, 1974, 132). To separate anything from nature was an absurd idea because the concept encompassed the entire universe. While today *physis* is ultimately the set of qualities commonly taken to embody nature, originally both *nómos* and *physis* were aspects of nature.

The Hellenistic Greeks also subscribed to the dualistic distinction between body and soul (Barbour, 1997). Whereas the human body was material and therefore a part of nature, the human soul was immortal and transcended the death of the body. This dichotomous conception of humans accentuated the difference between humanity and other creatures, which were viewed as lacking souls, and encouraged the perspective that humans are set apart from nature.

10.4.2 The Christian perspective

The Christian view of nature, dominating Western thought for hundreds of years, posits God as the creator of everything natural (Evernden, 1992, 20). Nature in this scheme is related to God, as is a work of art to the artisan. Anything 'unnatural' is not created by God and, therefore, has attached to it a stigma of immorality, while any phenomena or behavior classed as 'natural' has been given to humans by God and is wholly moral. The biblical story of the Garden of Eden provides an interesting twist to our relationship both to God and 'nature.' After being banished from the Garden, humans became the other, set in opposition to nature – a fundamental tenet of the nature–culture dualism.

The notion of the contingency of nature, as an expression of the divine will of God, is an important aspect of Christianity that laid the groundwork for modern science. Early empirical science emerged from the premise that the structure of the world could not be discerned through reflection on necessary first principles, as the Greeks had sought to do, but instead required active investigation to reveal the structure and order of God's creation, which are contingent upon God's will (Barbour, 1997). The notion that nature is a divine creation also grounded the assumption that order exists in the world and that this order is 'harmonious' in the sense that nature has an intelligible, coherent organization.

In contrast to the Greek perspective, Christian doctrine does not invoke a dualism between body and soul, but instead views mind, body and spirit as aspects of a psychosomatic unity (Barbour, 1997). This doctrine is best reflected in the resurrection account of Jesus Christ, which was a bodily resurrection, not just the 'escape' of an immortal soul from a mortal body. In the Christian perspective, humans do have a privileged position in the world relative to other creatures and objects, but this position is the result of the divine will of God bestowed upon unified persons, not of the distinctiveness of humans as creatures with separate bodies and souls.

10.4.3 Cartesian dualism

Renaissance views of nature inherited the Greek notion of distinctiveness between qualities of objects that are independent of humans and those that are dependent on human perception and interpretation (Soper, 1995, 53). Another important inherited idea was the role of human cognition in distinguishing these two types of qualities. Philosophers such as Copernicus (1473–1543), Telesio (1508–1588) and Bruno (1548–1600) contributed to the separation of humans from nature by arguing that nature could not be considered an organism and therefore has no intelligence or life of its own. Any order that did manifest itself in nature was attributable to something larger than or perhaps external to nature. These external forces were the 'laws of nature' or the processes that govern the movement of energy and matter throughout the physical world as in a machine (Cosgrove, 1990). Theologically, this idea was acceptable because God was, of course, the source and director of these universal laws (Tuan, 1968, 4).

Descartes (1590–1650) ushered dualism into the Modern period by arguing that while humans are born *Homo sapiens* as biological animals, our capacity to think, speak and create culture causes us to transcend the world of nature (Collingwood, 1945). Body was considered one substance, mind another (Gerber, 1997, 3). While Descartes acknowledged that humans had to be bound to nature, he could find no connection short of God, who inexplicably manifests Himself in the human pineal gland (Collingwood, 1945, 7). The material body of humans defines our base connection to worldly nature, while the human mind has the ability to transcend this base materiality. Descartes posited human thought and emotion as characteristics that make humans fundamentally different from material nature. Elaborating the views of Aristotle, he posited that this transcendence over nature includes all natural objects, such as animals who, in his view, have no intelligence, soul or capacity for emotion (Soper, 1995, 53). The effect of this Cartesian dualism was to restructure our sense of the world (Fitzsimmons, 1989, 109). Humanity was effectively severed once and for all from nature and placed above it, justified because nothing else in nature has intellect or soul. The mind and the works of the mind allow humanity to transcend the body of nature (Lease, 1995, 10).

Galileo (1564–1642), a contemporary of Descartes, described the world as having primary properties (real, mathematical) and secondary properties (human) (Evernden, 1992, 49). The distinction made by Galileo was partially a reaction against the science of Thomas Aquinas, who sought an understanding of natural processes that could best be described as wisdom (Evernden, 1992, 44). Nature was perceived by Galileo and others advocating Cartesian dualism as having no intellect or *purpose*; therefore, metaphysical questions about *why* nature operates as it does are not applicable to scientific inquiry. *How* was the appropriate question to be asked of nature, not *why*. This philosophy is strongly reflected in the scientific work of Sir Isaac Newton (1642–1727), who deliberately adopted a mechanistic perspective on the natural world. Most early scientists, including Newton, remained centred in the Church. Attempts to reconcile the mechanistic perspective with Christianity led to the view of God as the Divine Clockmaker, who wound up the clock of nature at creation, but now let His laws of nature operate without intervention (Barbour, 1997).

10.5 Contemporary distinctions

The metaphysical framework of Cartesian dualism underlies two main contemporary themes defining nature (Soper, 1995, 136). The first refers to the forms, processes, relations and causal powers operating in the biophysical realm, which have traditionally provided the objects of study for the physical or natural sciences. In this view, nature is synonymous with 'normal'. Forms in nature are permanent, recurring phenomena generated by environmental forces objectively perceived by humans from 'outside' the system. The second theme, the lay or surface discourse, refers to nature as the ordinary observable features or objects of the world such as animals, landforms, plant assemblages, and so on. This theme is the nature of immediate experience and aesthetic appreciation, what we see, feel, hear and touch in our interactions with the world around us. Both of these themes serve to objectify and externalize nature from human influence (Castree and Braun, 1998, 29). Latour (1993, 31–32) claims that these dualistic notions of nature have led to what he refers to as the three 'guarantees of modernity', which have mediated the scientific enterprise and other social relations with the biophysical environment since the Renaissance (Table 10.1).

Since the Western Rennaissance, the distinction between things 'human' and things 'natural' has dominated scientific conceptions of nature (Evernden, 1992). Contemporary natural science is based on the idea of nature as a realm apart from humans (Collingwood, 1945, 2). The separation between humans and the natural world gradually has become institutionalized via sociological practices and norms that have developed from the philosophical distinction between nature and culture (Evernden, 1992, 23). This conceptual dualism is so fundamental within the natural sciences that it is generally taken for granted (Burgess, 1978 and Smith and O'Keefe, 1980).

Defining nature as that which is not human places it in opposition to us. If 'an opposition ... between the natural and the human has been axiomatic to Western thought, and remains a presupposition of all its philosophical, scientific, moral and aesthetic discourse' (Soper, 1995, 38), it is not surprising that the underlying assumptions

Table 10.1 *The three guarantees of modernity.*

1.	It is not men who make Nature, Nature has always existed and has always already been there; we are only discovering its secrets.
2.	Human beings, and only human beings, are the ones who construct society and freely determine their own destiny.
3.	Nature and Society are absolutely separate.

Source: Latour, 1993, 31–32.

of science are infused with and affected by this perspective. Traditionally, our ideas about scientific objectivity have been built upon the distinction between nature and humans. We understand ourselves to be disconnected or set apart from other entities in universe. The Western tradition has also incrementally reinforced the notion that humans are, in a sense, *super*natural (Soper, 1995, 16). *Homo sapiens* has transcended nature in the sense that, unlike any other species, we are relatively independent of environmental or ecological constraints governing our existence. Consciousness, cognition, culture and spirituality distinguish us ontologically from other objects and beings of the biophysical realm. This view allows the phenomena of nature to be investigated by humans without the taint of human influence – a process referred to as the 'othering' of the biophysical world. An important implication of this perspective is that once the labor of humans does mix with nature, the phenomena altered by human interaction are no longer considered natural or are, at the very least, less natural (Johnson et al., 1997).

The traditional conception has led to separation of modes of intellectual inquiry within science at large. The physical and natural sciences, such as chemistry, physics and biology, investigate nature and natural processes (Borgmann, 1995, 33). Social sciences and the humanities explore the human and cultural realms: sociology, the arts and psychology (Lease, 1995, 10). Lastly, applied sciences, such as engineering and agriculture, examine what happens when the worlds of nature and culture collide (Jones, 1983), often directed specifically toward human modification of the world (Rhoads and Thorn, 1996). In geography, the three units – physical, social and applied – all coexist under one umbrella.

A further distinction, that between qualitative and quantitative forms of knowledge, derives from the philosophical and sociological distinctions of the traditional perspective. Quantitative knowledge, representing discreet ways of describing phenomena in mathematical terms, has traditionally been seen by Western science as reflecting universal truths existing in the world and, as such, has been synonymous with nature and physical science (Evernden, 1992, 49). According to this view, the natural system could only be known when the 'distorting effects of human projection' and subjectivity were removed from scientific inquiry (Evernden, 1992, 58).

Qualitative knowledge, used to describe phenomena which cannot easily be described in mathematical terms, has long been tied to human characteristics such as cognition, emotion and spirituality, the exploration of which is the domain of social science or the humanities (Cosgrove, 1990). The difference in focus (biophysical versus societal) and type of knowledge (quantitative versus qualitative) in the physical sciences versus the social sciences has led to the distinction of the 'two cultures'. The three interrelated dichotomies, nature–culture, physical science–social science and quantitative–qualitative, provide the bricks and mortar for the metaphorical wall that has been constructed between humans and the biophysical world. Any attempt at truly integrating humans into the biophysical must address these core conceptions.

10.6 Second nature: categorizing types of nature

Humans have dramatically changed biophysical environments through cultural modification of landscapes. In this sense, they themselves are now creators of new biophysical conditions. The concepts of 'humanity as creator' and 'nature as the world existing outside of humanity' come together in the notion of *second nature*. Second nature is a different kind of nature; it corresponds to landscapes inhabited and manipulated by human agency (Castree and Braun, 1998). Some have even gone so far as to label the simulated natures of television, magazines, GIS and theme parks as 'third nature,' one more step removed from the pristine ideal of non-human interference (Wark, 1994).

The term 'second nature' does more than simply refer to some object or set of objects; it points to a relationship between humans and the biophysical world (Evernden, 1992, 20). Secondary and tertiary categories of nature remain founded on the Cartesian distinction between humans and nature and, therefore, do not represent a fundamental departure from the traditional perspective. The introduction of these categories, however, does allow new questions to be explored. When nature is considered an 'other', it is objectified instead of being valued as participating in a relationship or dialectic with humans (Smith, 1984). Perceptions and behaviour cannot interact with an external nature-object without transforming it into second or even third nature. But an *objectified* nature cannot continue to exist after people have physically interacted with it. In a sense, then, the distinction between first nature and second nature should not be one of kind, but rather of degree (Smith and O'Keefe, 1980, 84). Banishing the categorical distinction should not be taken to mean that all forms of human perception and behaviour are equally appropriate (Castree and Braun, 1998, 4). It simply shifts the moral authority from a 'pristine nature' back to human society.

10.7 The end of nature: collapsing dualism via pragmatic considerations

Another variation on dualism acknowledges the pervasive influence of humans on earth's biophysical systems. This view focuses on the dilemma of how to separate humans from the biophysical environment in a world dominated by humans (Soule, 1995, 144 and Meyer, 1996, 217). Under these conditions it becomes difficult to locate, both theoretically and pragmatically, 'pristine' reference environments untainted by human influence. This perspective embraces dualism at the conceptual level, but argues that nature is now increasingly difficult to access because of widespread human effects. Nature/culture dualism collapses at the pragmatic level because we are witnessing 'the end of nature' (McKibben, 1989). Nature, as defined within the dualism framework, no longer exists because modification of it by humans has eliminated the basis for such a distinction, i.e. that which is unaffected by humans.

Whether humans have modified the earth system to the extent envisioned by 'End of Nature' proponents is debatable. Moreover, the perspective sidesteps an important issue – how do conceptions of nature inform ethics, which guide human interaction with the world? Is the end of nature a license for interacting with the biophysical environment however we see fit? The dissolution of dualism at the pragmatic level provides no 'deeper insight' (McKibben, 1989) on this issue.

At the pragmatic level, human culture is still seen as 'unnatural' on the grounds that it has the capacity to change the environment at greater rates, magnitudes and scales than populations of other organisms (Johnson et al., 1997). Although some species, referred to as keystone species, have a disproportionately large influence on ecosystems they inhabit (McIsaac and Brun, 1999), humans have seemingly become a noxious weed or invasive species at the global scale (Margulis, 1998). The scope of human-induced modifications to the biophysical world may be unusual within the context of the pace of most evolutionary and ecological changes, but that attribute in and of itself does not provide necessary and sufficient conditions to justify the claim that humans are *inherently* different from *all* other organisms – the basic premise of dualism (McIsaac and Brun, 1999).

10.8 Nature as a social construction: collapsing dualism via social epistemology

For some, the concept of nature is inherently a social construction developed to privilege the status of humans or groups of people (Soper, 1995, 75). Much has been written about different aspects of this conceptual reordering of the world from a variety of theoretical perspectives (Bird, 1987; Simmons, 1993; Harvey, 1993; Demeritt, 1994; Williams, 1994 and Cronon, 1995). A key aspect of this perspective is that it does not typically challenge the ontological distinction between humans and other phenomena in the world; instead, it focuses on the capacity of humans to have epistemic access to this distinction (Proctor, 1998). The view is founded in relativist philosophies that argue against the epistemic privilege of any particular version of knowledge (Rorty, 1979). Social factors tend to dominate science, and scientific knowledge itself is largely a social construct.

Because no version of knowledge about nature, scientific or otherwise, is inherently privileged, the concept of nature cannot be determined absolutely, but instead is mired in social debate among those adhering to different perspectives on nature. In other words, there is no such thing as *Nature*, there are only competing social visions of nature. Under this perspective, dualism largely collapses to monism at the epistemological level because social factors are viewed as ascendant in establishing the 'truth' about nature. A biophysical world may exist out there, but we do not have access to it ontologically in the sense that our epistemic practices will lead us to the one true human-independent understanding of this world.

The main argument levelled against constructivism is that the relationship between humans and nature is not only discursive but also material (Castree, 1997, 13). We must not confuse our ordering of the biophysical world with the world itself (Castree and

Braun, 1998, 19). Although the practice of science is social, it still revolves around interactions with a physical world (Castree and Braun, 1998, 26). Therefore, the production of scientific knowledge cannot be fully reduced to the social.

10.8.1 Naturalism: collapsing dualism via naturalized epistemology

Naturalized epistemology derives from the premise that philosophy cannot operate independently of the substantive content of scientific beliefs (Shapere, 1987). Philosophy as a way of knowing is not superior but, at most, equal to science. Thus, any attempt to explain the world philosophically must also account for the best available scientific knowledge, especially that of the natural sciences. This perspective provides the foundation for a monistic alternative to Cartesian dualism. Humans are recognized as an animal species, organically related to other creatures and to the inorganic world through the process of biological evolution. Humans are, literally, a part of nature – at least at the biological level (Gerber, 1997, 3 and Johnson et al., 1997). What we call 'the mind' and refer to as 'mental events' are epiphenomena that, in principle (although not yet in practice) can be reduced to neurophysiological processes in the brain, thereby eliminating the mind/body problem. The mind and its manifestations are nothing but matter and its functions.

If humans are indeed part of nature biologically and not a separate entity, we approach a position closer to naturalism and farther from humanism. The human condition, according to an evolutionary biological perspective, has developed through the imprinting of rational physical forces on the genetic legacy of the species (Peet, 1998, 13). Habitat and prospect-refuge theories are good examples of naturalistic approaches to explaining human preferences and outlooks on the world. The essence of these theories is that the sense of pleasure people feel in certain types of landscapes is genetically driven by those landscapes being favorable to biological survival (Appleton, 1975). The work of Appleton, for example, seeks to 'relate the idea of preference to a typology of landscapes through the medium of the biological, and more particularly the behavioral sciences' (Hudson, 1993). The preference we have for places with protected vistas, such as a promontory overlooking the ocean, is a genetically determined, biological impulse (Fisher, 1992). Human consciousness, thought and emotion are reduced to biological explanations and the human condition is re-defined as the human animal.

To completely collapse dualism, however, naturalism must also advance the notion that since humans are – first and foremost – biological animals, cultural processes can ultimately be reduced to biological or genetic imperatives (Smith, 1984, 19). Hence, the emergence of the field of sociobiology, whereby social behavior is reduced to explanations based on evolutionary principles (Wilson, 1975, 1978 and Dawkins, 1976). Here naturalism begins to encounter considerable difficulties. Few social scientists accept sociobiological explanations as adequately accounting for the complexity of the cultural milieu. One need only look to the risk involved in claiming social laws as natural laws – a political problem that has recurred throughout human history (Soper, 1995, 122 and Castree and Braun, 1998, 7) – to find evidence of the inadequacy of sociobiological explanations. Cultural evolution also differs fundamentally from biological evolution in that, while the latter involves unintentional natural selection operating on random mutations, the former consists of purposeful selection operating on intelligence-guided cultural variations (Rescher, 1996). Although brain structure and capacity developed through biological evolution, cognitive interaction with the world has evolved by cultural evolution – a process that seems to transcend purely biological explanations.

10.9 Multi-aspect monism: toward a complete collapse of dualism

The challenge remains to connect the social capacities of humans to the biophysical world without reducing nature to a cultural artifact (complete social construction of nature) or humans to a pure biological animal (biological naturalism). Scientific evidence leaves little room for doubting that humans are a species that has developed and evolved within the ecological context of a historical earth (Johnson et al., 1997, 582). In this sense, we are part of the earth ecosystem and cannot justify a 'supernatural' view of our origins relative to other organisms. The fact that *Homo sapiens* and chimpanzees have remarkably similar DNA most likely points to our common genetic heritage (Rolston, 1988, 33). Humans, however, are distinct from chimpanzees and other species in that we display a unique combination of physical and mental characteristics. This uniqueness does not imply any sort of hierarchical moral or spiritual superiority. After all, the same could be said for all species: each is both similar to and distinct from other animals (Tuan, 1968). Our consciousness and those processes generated by it, such as language and culture, are the qualities that most distinguish the human animal from other species. Whether consciousness is unique to humans or is only different in us as a matter of degree is an unanswered (and perhaps unanswerable) question.

What is needed to effect a complete collapse of dualism is not a *single-aspect monism*, in which mind is reduced to brain and culture to biology, but rather a *multi-aspect monism* that allows for the biological basis of humanity while, at the same time, acknowledging the distinguishing traits of humans, such as a level of consciousness capable of an unparalleled degree of self-awareness, including the cognizance of our own mortality (Polkinghorne, 1996, 60 and Barbour, 1997). The focus of this type of monism becomes levels of organization and the potential for emergence of high-level properties via multilevel process interactions within complex systems. Biological examples include the basic capacity for reproduction (which distinguishes the living from the non-living) and the development of brain capacity to the point where, from an evolutionary standpoint, the emergence of the self, or sentience, occurred in humans. Accompanying this development was the capacity for language, for developing and using tools, and for sophisticated social interaction which, in turn, led to the emergence of complex cultural practices, including contemporary science.

Consistent with contemporary scientific notions about the dynamics of complex, nonlinear systems, the behaviour of human beings (the system as a whole) cannot be reduced to a set of principles operating at the lowest level (e.g. brain mechanisms) because new principles of organization emerge (e.g. consciousness) and interact with low-level principles as complexity develops. This *processual systems perspective* reveals the profound shortcoming of Cartesian dualism with its focus on a substantive mind–brain distinction. Instead of mind versus brain, the emphasis is on mental activity versus neurophysiological mechanisms. According to the processual view, mental activity is not reducible purely to physiological mechanisms; instead, these two phenomena are coordinated manifestations of a complex, unified process (Rescher, 1996). Thus, humans represent highly complex organisms set within nature who have developed distinctive cognitive capabilities that differentiate us from other entities in the natural world. Multi-aspect monism not only embeds mental activity and brain function within a unified ontological framework, it proposes that through the natural and social sciences we are acquiring epistemic access to this unified framework. Conceptually, the *content* of mental processes cannot be reduced merely to neurophysiological processes because these two types of processes are different, but interactive components of a multi-aspect holistic complexity we refer to as a person. Moreover, interaction among humans at the experiential level leads to emergence of an even higher level of organization – cultural processes, which comprise, but also influence, the experiences of individuals.

Eliminating the traditional distinction between mind and body, or nature and culture, typified in Cartesian dualism and replacing it with a naturalistic view of embodied consciousness grounded in physical experience, shifts the focus from nature as the 'other' to nature as the process or experience in which humans are embedded. This view situates people within the physicality of the world (Spurling, 1977). When we effect change in nature, we affect not only ourselves but also nature itself because it can be seen as an extension of ourselves (Evernden, 1992, 101). Thus the concept of nature as a reality characterized by immutable order is replaced by the concept of nature as a dynamic, continuously evolving system that changes as the history of human interaction with the biophysical environment progresses.

10.10 Nature–society relations in geography: a tradition of dualism

Contemporary geography, even though it includes practitioners that straddle the social and physical (natural) sciences, is deeply entrenched in the tradition of Cartesian dualism and its separation of nature and society. Although the discipline has been characterized as one in which the 'two cultures' of the human and natural sciences successfully have merged with one another (Ziman, 2000, 181), reality undermines this characterization. The two sides of the discipline largely operate independently of one another and the increasing degree of specialization in human and physical geography is only increasing the gulf between them (Trimble, 1992). Since physical geography is a composite science, in part grounded within principles of physics, biology and chemistry, most physical geographers subscribe to the concept of nature as defined by the natural sciences

(Osterkamp and Hupp, 1996 and Katz, 1997). A clear distinction is made between culture and those natural environmental characteristics loosely referred to as 'nature' (Katz and Kirby, 1991, 262). The domain of physical geography is the biophysical world. If humans are considered, it is only the effects of human action on the biophysical world that are of interest, not the motivations behind the effects. Although this stance has thwarted interaction with human geography, an extreme position that denies knowledge of a biophysical reality independent of human perception and consciousness is an unsatisfactory solution for physical geography, whose primary focus remains the material reality of biophysical systems (Burgess, 1978, 71; Liu, 1989 and McKibben, 1989).

Human geographers are still smarting intellectually from geography's incipient foray into environmental determinism and social Darwinism (Peet, 1998, 13) in the early 20th century. This naturalistic outlook promoted the view that all aspects of human behaviour and genetic character can be explained by the influence of the biophysical environment:

> *Man [sic] is a product of the earth's surface. This means not only that he is child of the earth, dust of her dust, but that the earth has mothered him, fed him, set him tasks, directed his thoughts, confronted him with difficulties that have strengthened his body and sharpened his wits, given him his problems of navigation or irrigation, and at the same time whispered hints for their solution. She has entered into his bone and tissue, into his mind and soul.*
>
> Semple, 1911, 1

Environmental determinism eventually degenerated into scientific racism (Peet, 1998, 14) by using environmental characteristics and their effect upon human development to affirm the superiority of northern Europeans and the inferiority of equatorial peoples (Johnston, 1997, 43). The justified strong backlash against this misguided approach to explaining human culture has haunted human geography for over 100 years by proscribing *any* explanation of cultural activities that invokes a human response to the biophysical environment. Thus human geographers have implicitly adopted dualism by seeing the biophysical environment as an irrelevant domain for addressing social issues and by advancing explanations that invoke human processes only.

The rising prominence of environmental issues and concerns in society and the position of geography at the interface between the natural and social sciences implies that geographers should figure prominently in the dissolution of the human-nature dichotomy (Graf et al., 1980, 281 and Gerber, 1997, 14). However, the current lack of interaction between human and physical geography only reinforces the conception that humans are alienated from nature and limits our vision of the full range of forces, both biophysical and human, effecting change in the physical environment (Fitzsimmons, 1989 and Soper, 1995). It has affected the ways in which geographers view landscape change by separating the dynamics of human agency from the biophysical effects of that agency (Jones, 1983, 430). Given that humans increasingly are changing and responding to changes in the biophysical environment (Hooke, 1994 and Roe, 1996), the validity of the split between human and physical geography must be challenged (Rhoads, 1999). Most landscapes inhabited by *Homo sapiens* are 'mutually determined' through the complex interaction of biophysical and human processes (Cronon, 1983, 13). The time is ripe to call into question the basic assumptions on which the traditional split between human and physical geography is grounded. Human and physical geographers should embark on a new era of collaboration to unravel the complexity of human–biophysical interactions. They must challenge the heritage of dualism and strive to transcend the intellectual constraints it has placed on geography.

10.11 Stream naturalization: an integrated geographic research theme

Watershed science provides a superb research context for integrating human and physical geography. The current focus of watershed science emphasizes integration of human and biophysical elements to provide a knowledge base to guide watershed management (National Research Council, 1999). Approaches that transcend the bounds imposed on science by the 'two cultures' are needed to examine the complexity of watershed problems.

The concept of stream naturalization provides an example of an incipient attempt to integrate human and physical geography to address research questions related to interaction between human and biophysical factors within the context of watershed management. Over the past several decades, a new attitude has emerged concerning the management of streams and rivers throughout the United States. This attitude is characterized by multi-objective rather than single-objective management, with a strong emphasis on the need to consider environmental quality in the management process. Conventional concepts related to environmental management of fluvial systems affected by human modification include stream restoration, rehabilitation and enhancement (National Research Council, 1992) where restoration refers to a return of the system of interest to a pre-disturbance state, rehabilitation refers to a partial return to the pre-disturbance state, and enhancement refers to any improvement of ecosystem function. The first two concepts use the pre-modified condition as the reference state for management. Reference identification for enhancement is less clear, but examples provided by the National Research Council (1992) suggest that it relates loosely to the pre-disturbance condition and relies on expert judgment by ecologists as to what constitutes improvement.

All three concepts proposed by the NRC have shortcomings when applied to human-dominated environments, such as agricultural and urban landscapes. In these types of settings, the biophysical environment has been radically modified and reference conditions representative of the pristine state often do not exist. Moreover, scientific information on pre-disturbance conditions commonly is unavailable because major landscape modification occurred prior to the systematic collection of such information.

Even if adequate information is available and restoration is economically acceptable, which typically is not the case, changes in land use at the watershed scale in most cases have been so extensive that the pristine state of streams is no longer sustainable under current conditions. Finally, restoration, rehabilitation and enhancement are value-laden terms that inherently embrace dualism by privileging the 'pristine', or human-absent, state of ecosystems as the reference condition for ecological 'integrity' or 'health'. This perspective separates humans from the 'natural' world and views all human intervention as 'disturbance' of natural conditions (Graf, 1996, 2001). What it fails to realize is that in certain environments, humans and human culture now dominate the landscape and will continue to do so for the foreseeable future, regardless of what efforts are made to 'improve' environmental quality. An environment dominated by humans is now the natural condition of the landscape and reference to the 'pristine' state has little relevance for management of this landscape.

Despite the irrelevance of the pristine state as reference condition, the emergence of community-based initiatives to 'improve' environmental quality of human-dominated landscapes is indicative of an inherent desire to 'naturalize' these landscapes. The content of this desire for naturalization has yet to be explored in detail, but interaction with agricultural and urban stakeholders in Illinois suggests that naturalization generally reflects an aspiration to undo much of the human-induced homogenization of streams that has occurred in the past (Rhoads and Herricks, 1996). Thus the reference condition becomes the current simplified state of the system, and preferences for naturalization reflect a desire to move the system away from this state toward one with greater complexity and diversity in terms of hydraulic, geomorphologic and ecologic characteristics (Rhoads et al., 1999). Acceptable strategies for achieving this goal seem to vary from community to community. Thus specific conceptions of naturalization are suitably place-based.

In sum, the concept of stream naturalization recognizes the growing community-level interest in the establishment of sustainable, morphologically and hydraulically varied, yet dynamically stable, fluvial systems that are capable of supporting healthy, biologically diverse aquatic ecosystems; embraces the notion that particular conceptions of 'natural' are community-based and place-specific; and acknowledges that recurring human interaction with biophysical components of fluvial systems is often part of the contemporary and future natural environment in resource-rich settings. The underlying basis for this concept is a multi-aspect monism that embraces the embeddedness of humans in a real nature, the resulting spatial complexity of nature and the evolution of nature through interaction among biophysical and cultural processes. Humans may be a mammalian weed (Margulis, 1998), but weeds are not 'supernatural' and, like any other weed, we can alter ecosystems by over-consuming resources, destroying habitat and becoming an invasive species. In contrast to other organisms, including weeds, our self-awareness allows us to recognize the consequences of our actions on ourselves and on other phenomena and to thereby consider, through moral deliberation, changing how we interact with the non-human world.

On a pragmatic level, the concept of stream naturalization serves as a keystone for a research program focusing on watershed management in agricultural and urban landscapes of the Midwest that integrates physical and human geography (Rhoads et al., 1999 and Wade et al., 2002). This program is based on the premises that environmental management of streams in these human-dominated landscapes is a process that is fundamentally social in nature and that attempts to naturalize streams have been limited not only by social factors, but by incomplete understanding of the connections among hydraulic, geomorphological and ecological processes in human-modified streams. Key components of the research involve developing an understanding of the local place-based socio-cultural contexts within which decisions about stream naturalization are made, developing an improved base of geomorphological knowledge on the dynamics of human-modified streams to support local decision-making, and exploring how scientific information on stream geomorphology shapes community visions and influences decision-making about stream naturalization. The first component is the focus of human geography, whereas the second is the focus of physical geography. Integration of human and physical geography occurs via the third component, whereby geomorphological information is presented to local communities and the extent to which this information is absorbed and acted upon by these communities becomes a focus of human-geographical analysis. This research program is described, not to provide a definitive model for integration of human and physical geography, but to provide an inceptive example for moving toward the goal of an integrated geography.

10.12 Conclusion

This chapter has addressed the meaning of the concept of nature within Western thought and the influence of this meaning on science at large and on geography specifically. The predominant perspective has been to associate nature with the biophysical (non-human) world and to view humans as set apart or separate from nature. Modern geography largely subscribes to this perspective, which partly accounts for the disjointed relationship between human and physical geography. The distinction between the natural and social sciences has evolved from nature–society dualism. The desire of human geography to be seen as a legitimate social science and of physical geography to be seen as a legitimate natural science sustains a dichotomous geography (Rhoads, 1999). Breaking down this entrenched dualism seems to be a prerequisite to an integrated geography. Alternative perspectives on nature, such as multi-aspect monism, provide possible starting points for developing a conceptual foundation for an integrated geography.

The community of geographers must decide whether integration of the human and physical sides of the discipline is a desirable goal. We suspect, however, that geography's broad educational scope, and its propaganda about being an 'integrative' field of knowledge, draw many to the discipline. Thus there appears to be real motivation for fulfilling the promise of true integration. It would, however, be foolhardy to advocate that a new generation of geographers should be trained as 'integration' generalists. Each sub-field of geography increasingly requires mastery of a considerable body of specialized knowledge to achieve proficiency. Moreover, in this age of expanding information, information technology and transdisciplinary problems, it is beyond the scope of individuals to accomplish integrated research (Ziman, 2000). Instead, integration must become a shared commitment, one that the geographic community at large sees as a central intellectual focus of the discipline. Physical geographers must decide not only if they are comfortable at home in a discipline that includes human geography colleagues but, more importantly, if they want actively to engage in collaborative research with these colleagues. Some of us can continue to specialize without interaction, but a move toward an integrated geography will require cooperation with human geographers by at least a small contingent of physical geographers.

References

Angeles, Peter A. 1992: *The Harper Collins Dictionary of Philosophy,* 2nd ed. Harper Collins, New York.

Appleton, Jay 1975: *The Experience of Landscape.* John Wiley and Sons, New York.

Barbour, I. 1997: *Religion and Science: Historical and Contemporary Issues.* Harper, San Francisco.

Bate, Jonathan, 1991: *Romantic Ecology.* Routledge: London.

Bauer, Bernard O. 1996: 'Geomorphology, geography, and science.' In B.L. Rhoads and C.E. Thorn (eds.) *The Scientific Nature of Geomorphology,* Wiley, New York, 381–414.

Bird, Elizabeth Ann R. 1987: 'The social construction of nature: theoretical approaches to the history of environmental problems'. *Environmental Review* 11 (4), Winter: 255–64.

Borgmann, Albert 1995: 'The nature of reality and the reality of nature'. In M.E. Soule and G. Lease (eds.) *Reinventing Nature: Responses to Postmodern Deconstruction.* Island Press, Covelo, CA.

Bowler, Peter J. 1993: 'Science and the environment: new agendas for the history of science?' In M. Shortland (ed.) *Science and Nature: Essays in the History of the Environmental Sciences.* British Society for the History of Science, London.

Burgess, Rod 1978: 'The concept of nature in geography and marxism'. *Antipode* 10 (2), 1–11.

Castree, N. 1997: 'Nature, economy and the cultural politics of theory: "the war against the seals" in the Bering Sea, 1870–1911'. *Geoforum* 28: 1–20.

Castree, Noel, and Bruce Braun 1998: 'The construction of nature and the nature of construction: analytical and political tools for building survivable futures'. In B. Braun and N. Castree (eds.) *Remaking Reality: Nature at the Millennium.* Routledge, New York.

Chorley, R.J. 1978: 'Basis for theory in geomorphology'. In C. Embleton, D. Brunsden, and D.K.C. Jones, *Geomorphology: Present Problems and Future Prospects.* Oxford University Press, Oxford.

Collingwood, R.G. 1945: *The Idea of Nature.* Oxford University Press, New York.

Colwell, Tom. 1987: 'The ethics of being part of nature'. *Environmental Ethics* 9, 99–113.

Cosgrove, Denis 1990: 'Environmental thought and action: pre-modern and post-modern'. *Transactions of the Institute of British Geographers* 15, 344–58.

Cronon, William 1983: *Changes in the Land: Indians, Colonists, and the Ecology of New England.* Hill and Wang, New York.

Cronon, W. 1995: 'The trouble with wilderness'. *New York Times* 42–43.

Dawkins, R. 1976: *The Selfish Gene.* Oxford University Press, Oxford.

Demeritt, D. 1994: 'The nature of metaphors in cultural geography and environmental history'. *Progress in Human Geography* 18: 163–85.

Evernden, Neil 1992: *The Social Creation of Nature.* Johns Hopkins, Baltimore.

Fisher, Bonnie S. 1992: 'Fear of crime in relation to three exterior site features: prospect, refuge and escape'. *Environment and Behavior* 24(1), 35–65.

Fitzsimmons, Margaret 1989: 'The matter of nature'. *Antipode* 21 (2), 106–20.

Gerber, Judith 1997: 'Beyond dualism – the social construction of nature and the natural "and" social construction of human beings'. *Progress in Human Geography* 21 (1), 1–17.

Gifford, Terry 1996: 'The social construction of nature'. *Interdisciplinary Studies in Literature and Environment* 3, 27–36.

Glacken, Clarence J. 1967: *Traces on the Rhodian Shore: Nature and Culture in Western Thought from Ancient Times to the End of the Eighteenth Century.* University of California Press, Berkeley, CA.

Gober, Patricia 2000: 'In search of synthesis'. *Annals of the Association of American Geographers* 90 (1), 1–11.

Graf, William L. 1996: 'Geomorphology and policy for restoration of impounded American rivers: what is "natural?"'. In B.L. Rhoads and C.E. Thorn (eds.) *The Scientific Nature of Geomorphology.* John Wiley and Sons, New York, 443–475.

— 2001: 'Damage control: restoring the physical integrity of America's rivers'. *Annals of the Association of American Geographers* 91 (1), 1–27.

Graf, W.L., Trimble, S.W., Toy, T.J. and Costa, J.E. 1980: 'Geographic geomorphology in the eighties'. *The Professional Geographer* 32, 279–84.

Gregory, K.J. 2000: *The Changing Nature of Physical Geography.* Arnold, London.

Haraway, Donna 1994: 'A game of cat's cradle: science studies, feminist theory, cultural studies'. *Configurations* 1, 59–71.

Harvey, D. 1993: 'The nature of the environment: the dialectics of social and environmental change'. *Socialist Register* 30, 1–51.

Harvey, David 1996: *Justice, Nature and the Geography of Difference.* Blackwell, Cambridge, MA.

Hooke, Roger LeB 1994: 'On the efficacy of humans as geomorphic agents'. *GSA Today* 4(9), September, 217, 224–225.

Hudson, Brian J. 1993: 'Bennett's five towns: a prospect-refuge analysis'. *British Journal of Aesthetics* 33(1), 41–51.

Johnson, D.L., Ambrose, S.H. Bassett, T.J., Bowen, M.L., Crummey, D.E., Issacson, J.S., Johnson, D.N., Lamb, P., Saul, M. and Winter-Nelson, A.E. 1997: 'Meanings of environmental terms'. *Journal of Environmental Quality* 26 (3), 581–89.

Johnston, R.J. 1997: *Geography and Geographers: Anglo-American Human Geography since 1945,* 5th ed., Arnold, London.

Jones, David K.C. 1983: 'Environments of concern'. *Transactions of the Institute of British Geographers* 8, 429–57.

Katz, Cindi and Kirby, Andrew 1991: 'In the nature of things: the environment and everyday life'. *Transactions of the Institute of British Geographers* 16, 259–71.

Katz, Eric 1997: 'Nature's presence: reflections on healing and domination'. In A. Light and J.M. Smith (eds.), *Space, Place and Environmental Ethics*. Rowman and Littlefield, New York.

Latour, B. 1993: *We Have Never Been Modern*. Harvard University Press, Cambridge, MA.

Lease, Gary 1995: 'Nature under fire'. In M.E. Soule and G. Lease (eds.), *Reinventing Nature: Responses to Postmodern Deconstruction*. Island Press, Covelo, CA.

Liu, Alan 1989: *Wordsworth: The Sense of History*. Stanford University Press, Stanford.

Margulis, L. 1989: *Symbiotic Planet*, Basic Books, New York.

McIsaac, G.F., and Brun, M. 1999: 'Natural environment and human culture: defining terms and understanding worldviews'. *Journal of Environmental Quality* 28, 1–24.

McKibben, B. 1989: *The End of Nature*. Random House, New York.

Meyer, William B. 1996: *Human Impact on the Earth*. Cambridge University Press, Cambridge.

National Research Council 1999: *New Strategies for America's Watersheds*. National Academy Press, Washington D.C.

National Research Council 1992: *Restoration of Aquatic Ecosystems*. National Academy Press, Washington D.C.

Osterkamp, Waite R. and Hupp, Cliff R. 1996: 'The evolution of geomorphology, ecology, and other composite sciences'. In B.L. Rhoads and C.E. Thorn (eds.) *The Scientific Nature of Geomorphology*. Wiley, New York.

Peet, Richard 1998: *Modern Geographical Thought*. Blackwell, Oxford.

Polkinghorne, John 1996: *Beyond Science: The Wider Human Context*. Cambridge University Press, New York.

Proctor, J.D. 1998: 'The social construction of nature: relativist accusations, pragmatist and critical realist responses'. *Annals of the Association of American Geographers* 88, 352–376.

Rescher, N. 1996: *Process Metaphysics*. State University of New York Press, Albany, NY.

Rhoads, Bruce L. 1999: 'Beyond pragmatism: the value of philosophical discourse for physical geography'. *Annals of the Association of American Geographers* 89 (4), 760–70.

Rhoads, Bruce L. and Thorn, Colin E. 1993: 'Geomorphology as science: the role of theory'. *Geomorphology* 6, 287–307.

— 1994: 'Contemporary philosophical perspectives on physical geography with emphasis on geomorphology'. *Geographical Review* 84 (January), 90–101.

— 1996: 'Observation in geomorphology'. In *The Scientific Nature of Geomorphology*. B.L. Rhoads and C.E. Thorn (eds.), Wiley, New York, 21–56.

Rhoads, Bruce L., Wilson, David, Urban, Michael, and Herricks, Edwin E. 1999: 'Interaction between scientists and nonscientists in community-based watershed management: emergence of the concept of stream naturalization.' *Environmental Management* 24 (3), 297–308.

Rhoads, B.L. and Herricks, E.E. 1996: 'Naturalization of headwater agricultural streams in Illinois: challenges and possibilities.' In A. Brookes and F.D. Shields (eds.) *River Channel Restoration: Guiding Principles for Sustainable Projects*. John Wiley and Sons, Chichester.

Richards, Stewart 1987: *Philosophy and Sociology of Science,* 2nd ed. Basil Blackwell, Oxford.

Roe, Emery 1996: 'Why ecosystem management can't work without social science: an example from the California northern spotted owl controversy'. *Environmental Management* 20 (5), 667–74.

Rolston III, Holmes 1988: *Environmental Ethics: Duties to and Values in the Natural World.* Temple University Press, Philadelphia.

Rorty, R. 1979: *Philosophy and the Mirror of Nature.* Princeton University Press, Princeton, NJ.

Schumm, Stanley A. 1991: *To Interpret the Earth: Ten Ways to be Wrong.* Cambridge University Press, New York.

Shapere, D. 1987: 'Method in the philosophy of science and epistemology'. In *the Process of Science.* N.J. Nersessian (ed.), Martinus Nijhoff, Dordrecht, 1–39.

Sherman, Douglas J. 1996: 'Fashion in geomorphology'. In B.L. Rhoads and C.E. Thorn (eds.), *The Scientific Nature of Geomorphology: Proceedings of the 27th Binghamton Symposium in Geomorphology held 27–29 September 1996,* Wiley, New York, 87–114.

Simmons, I.G. 1993: *Environmental History: A Concise Introduction.* Blackwell, Cambridge, MA.

Smith, Neil 1984: *Uneven Development: Nature, Capital and the Production of Space.* Blackwell, Oxford.

Smith, Neil and O'Keefe, Phil 1980: 'Geography, Marx and the concept of nature'. *Antipode* 12 (2), 30–39.

Soper, Kate 1995: *What is Nature?* Blackwell, Oxford.

Soule, Michael E. 1995: 'The social siege of nature'. In M.E. Soule and G. Lease (eds.) *Reinventing Nature: Responses to Postmodern Deconstruction,* Island Press, Covelo, CA, 137–170.

Spurling, Laurie 1977: *Phenomenology and the Social World: The Phenomenology of Merleau-Ponty and its Relation to the Social Sciences.* Routledge, Boston.

Trimble, Stanley W. 1992: 'Preface'. In L.M. Dilsaver and Colten C.E. (eds.) *The American Environment: Interpretations of Past Geographies.* Rowman and Littlefield, Tontowa, NJ.

Tuan, Yi Fu 1968: *The Hydrologic Cycle and the Wisdom of God: A Theme in GeoTeleology.* University of Toronto Department of Geography Research Publications, Toronto.

Tuan, Yi-Fu 1974: *Topophilia: A Study of Environmental Perception, Attitudes and Values.* Prentice-Hall, Engelwood Cliffs, NJ.

Vicedo, Marga 1995: 'Scientific styles: toward some common ground in the history, philosophy, and sociology of science'. *Perspectives on Science* 3 (2), 231–54.

Wade, R.J., Rhoads, B.L., Rodríguez, J., Newell, M.D., Wilson, D., Herricks, E.E., Bombardelli, F., Garcia, M.H. and Schwartz, J. 2002: 'Integrating science and technology to support stream naturalization near Chicago, Illinois', *Journal of the American Water Resources Association,* 38, 931–944.

Wark, M 1994: 'Third nature'. *Cultural Studies* 8, 115–32.

Williams, M. 1994: 'The relations of environmental history and historical geography'. *Historical Geography* 20, 1.

Williams, Raymond 1980: *Problems in Materialism and Culture.* Verso, London.

Wilson, E.O. 1975: *Sociology: The New Synthesis.* Harvard University Press, Cambridge.

Wilson, E.O. 1978: *On Human Nature.* Harvard University Press, Cambridge.

Ziman, J. 2000: *Real Science.* Cambridge University Press, Cambridge.

11
Ethical grounds for an integrated geography

Keith Richards

11.1 The integration ethic in geography

Over the years, there have been several attempts to integrate the dual concerns of geography with the natural environment and with human society, and these have been both substantive and methodological in their focus. The former, for example, have included the French *Annales* school, forms of areal differentiation, the landscape school, and various types of human ecology. In general, these have been richly descriptive approaches which combine detailed empirical, observational study with a persuasive literary skill (Haggett, 1967), and which represent the natural and cultural landscapes as mutually determined, and as reflected in their spatial co-association. Variants on the theme of causality in these frameworks have ranged from environmental determinism, emphasizing the causal power of nature, to possibilism, in which nature offers options from which humans select. Notwithstanding this history, the case for integration has not always been viewed favourably, and the nature of geography has even been discussed with scant regard for its concerns with the natural world; Dear (1988), for example, makes virtually no reference to a 'physical' geography, let alone its elements. Indeed, Johnston (1983) has argued that, although issues of resource management might appear to be a suitable theme for integrated analysis, different specialists generally focus on particular sub-fields (resource appraisal, resource allocation, etc.), and in doing so use academic language, analysis and discourse in relation to their own field, and vernacular language and description to represent the others (Johnston, 1986). This is a common failing of what is often claimed to be interdisciplinary reseach, and afflicts geography internally as much as it affects conversations between disciplines. However, there has been a millennial reawakening of interest in dialogue within geography (Massey, 1999), so it is timely to revisit these debates.

Methodological as well as substantive grounds to justify integration have also been considered, albeit with limited success. For example, a search for a natural, common, geographical scale, at some rather indeterminate, but intermediate level, is undermined by contemporary appeals to processes acting across the global–regional–local scale range. Emphasis has also been placed on commonality of methods and techniques of analysis, such as quantitative spatial analysis, or even philosophies, such as realism. However, none of these provides unique grounds for a specifically geographical unity, although Chapman

(1977) demonstrated the potential of combining a geographical spatial scale with systems analysis. An alternative approach followed here involves searching for ethical grounds for an integrated 'moral geography' (Smith, 2000). This debates a theoretical ethical framework for valuing *both* the environmental and the social, and results in a form of human ecology. Discovering ethical value in the environment often seems to require an anthropocentric viewpoint, in which that value is expressed only when there is a danger of harm to dependent moral actors (that is, humans). However, an argument may be developed in which it is possible to map certain structural and emergent properties of ecosystems onto similar aspects of social relations, in ways that suggest a common conception of rights, entitlements and obligations. This ethical framework is defined not just from the standpoint of the obligations of human agents, but from the existence of networks of 'real' structures and processes that give both nature and society their co-evolved emergent properties of mutualism. This suggests a basis for the uncovering of 'real' ethical value in the natural world, which by this argument lies less in the observable entities (the pandas and tigers of conservation politics), than in the unobservable relations between those entities that give the world a degree of stable structure.

This general argument is supported below by a case study drawing on integrated geographical knowledge of networks of environmental and social processes. This concerns water resource development in the Awash Valley in Ethiopia, where various environmental and social networks of rights, entitlements and obligations, underpinned by established practices (processes) of range management and economic exchange, have been disrupted by an abrupt, externally imposed and non-participatory development process. The consequences of this betoken actions that may be construed as having been unethical. This case study concerns the human ecology of a problem in which humans are closely and directly dependent on natural environmental and renewable resources. A potential critique arising from this is that the argument for an ethical integrated geography relies on the specific exemplary cases being of such a kind, and that the argument is therefore not general. However, the counter to this is that such cases show clearly the interdependence of humans and their environment at a particular and appropriate scale, that this confirms the fundamental nature of the relationship and the ethical framework through which it is sustained, and that other linkages of this kind occur in other social circumstances at different scales. This rests on the observation that humans are inseparable from nature, and that the *fundamental* values which underpin this are independent of the *degree* to which the relationship may be apparent in any particular time and place, and its particular manifestation. It is then the task of an integrated geographical ethic to identify and analyse these relationships.

11.2 Grounded ethics as grounds for integration

To establish ethical grounds for an integrated geography, it may seem simplest to begin from an explicitly anthropocentric position, without immediate regard for the environment. However, this guarantees neither that the outcome will ensure the ethical standing of the environment itself, nor that it will support the integration of geography. Thus a more helpful, if also a more challenging, approach is to attempt to begin with ethical value in the environment itself. Attempts to define such ethical value may suggest the practice of assigning a monetary metric, but this should be resisted as a starting point, since it implies an immediate category mistake in which ethical values are replaced by consumer preferences (Sagoff, 1988). Further, whilst '. . . environmental ethics might . . . be conceived as a type of lived ethics' (Peterson, 2001, 2), implying the exercise of grounded, practical values, a problem for environmental ethics is that it seems powerless to overcome the pervasive effect of utilitarian individualism, now deeply enshrined in political and economic institutions. Thus, something is required beyond an equation of value with cost, and which addresses questions of moral action: hence, the need to consider whether there are intrinsic ethical grounds for defining environmental value. There are clearly no simple ways of achieving this. Although some may claim that there are real environmental values in the natural world, O'Neill (1997) argues that it is extremely difficult to establish what these might be, beyond reliance on acts of faith. Ironically, this implies that most attempts to define rationally the values on which environmental ethics are based, appear necessarily to be anthropocentric to some extent, even if the ultimate goal is to uncover the independent ethical value of the environment. In any event, anthropocentrism seems inherently to follow from the argument that only human agents have the capacity for behaviour and action based on reason and moral judgement, and the ability to establish institutions to sustain and develop that judgement. Candidates for this anthropocentric starting point, then, include utilitarian, rights-based and obligations-based approaches. The following discussion of these approaches leads eventually to the conclusion that networks of reciprocal relationships in nature and society represent real underlying structures that do have ethical value, and that this provides a basis for valuing the environment independently of, but in common with, human rights and obligations.

11.2.1 Obligations and ethics

Some forms of utilitarianism allow moral standing, at least in non-human animals, on the grounds of their capacity to suffer. However, this does not protect plants, landscapes and abstract environmental phenomena (ecosystems, biodiversity), nor does it avoid the kinds of trade-off that have to be reduced to cost–benefit analysis. Thus these are unreliable bases for determining a broad and sustainable definition of environmental value. One alternative anthropocentric approach, however, focuses on *rights*. These may be limited to claims about *fundamental* rights (the rights to food and shelter), although these often seem 'thin' compared to the rich, culturally varied entitlements that may be assumed, claimed, established by evolved practice or defined by institutions that regulate rights established primarily by consensus. Claims to rights, however, do often descend into rhetoric that cannot be substantiated and may be highly inflationary (Blackburn, 2001), largely for the same reasons that real environmental values seem so elusive. Thus O'Neill (1997) concludes that an alternative focus is required, on *obligations* or duties, because it is in obligations that the capacity for action resides. Rights appear to have little significance in the absence of counterpart obligations, and it is the obligation, rather than the right, that is fundamental: rights can only be claimed, whereas obligations are discharged. If it can be demonstrated that non-human animals and abstract environmental entities may be the subject of fundamental human obligations, even though they cannot always be said to have rights in themselves, this at least defines an ethical position which is protective of the environment, and which can underpin institutions whose function is to ground, exercize and protect these obligations.

Following Kant, fundamental obligations that determine ethical value in the environment may be identified through a search for universalizable principles (O'Neill, 1997). Human obligations must be *universal* to be the source of *fundamental* ethical principles, because logically they must at least apply to *all* humans. It cannot, for example, be a universal ethical obligation to injure others, because the act of doing so may prevent other humans from fulfilling this very obligation. It can, however, be a universal obligation to *refrain* from injuring others; and this obligation also matches a *right* not to be injured. There will be circumstances in which an injury, having been inflicted, may be adjudged to have been acceptable, but the judgement about this relies on the establishment of institutional structures (such as, for example, a code of law). Such institutionalized values may not be fixed, since they are likely to have to adapt to changing social and cultural norms. However, once a fundamental obligation not to injure is established, it has the potential to provide protection for the environment as well as humans. O'Neill (1997, 137), for example, concludes that '. . . if the rejection of systematic or gratuitous injury to other agents is a fundamental obligation, then it will also be obligatory not to damage or degrade the underlying powers of renewal or regeneration of the natural world'. Thus human obligations can be extended to non-human and abstract features of the world, such as landscapes, ecosystems and biodiversity, primarily because damage or injury to nature can indirectly cause damage or injury to fellow humans. We (humans) have established institutions – from indigenous common property resource management institutions, through environmental protection agencies, to the Inter-Governmental Panel on Climate Change – whose task it is to seek to regulate and mitigate deliberate and unintended damaging impacts on the environment, which may injure others and, indeed, ourselves. It is when those institutions fail that environmental degradation occurs, and those with the weakest political voice then suffer injury accordingly, in the form of loss of resource and increasing poverty (Dasgupta et al., 2000).

For some, this basis for environmental ethics may nevertheless seem unpalatable because of its anthropocentrism; it does not value the environment intrinsically, but only insofar as damage by humans to the environment may injure other human agents. However, since it appears that only human agents have the power to establish institutions to regulate the obligation not to injure (and, indeed, only humans appear to have the capacity to determine values), *initially* this may seem inevitable. Institutions are necessary to regulate the maintenance of environmental (and other) values, because there are always likely to be trade-offs to be made when environmental attributes are differentially treated as resources by various groups or individuals, so that there are potentially those who benefit and those who do not. Institutions that regulate behaviour in respect of the environment derive from similar bases to those involved in the ethics of human inter-relationships, and are therefore a matter of common law as much as economics. Decisions about the legitimacy of harm may thus be achieved more fundamentally using political or legal instruments (for example, through a democratic process or using a juridical process such as an environmental or citizens' jury; Sagoff, 1998 and O'Connor, 2000) than by appealing to economic procedures such as hedonic pricing or contingent valuation that attempt to assign monetary value (Pearce et al., 1989, 51–81), although such methods may have practical utility in specific contexts. Regulation by ethically based institutions rather than by economic criteria ensures that the interests of the powerful are limited, which is essential when many environmental phenomena or attributes are forms of common property.

To justify environmental protection using an anthropocentric principle of the avoidance of harm may be criticized in two ways (Baxter, 1999). The first challenge is the counter-argument that reliance on the establishment of universalizable principles may not always yield environmental protection, since even when such principles are accepted by humans, this does not guarantee that the obligations thereby created will be discharged fairly. However, this criticism fails to distinguish between *fundamental* moral obligations and institutional negotiation to *regulate* these obligations. Institutions that exist to regulate human action may or may not be based on fundamental ethical principles. However, those that *are* based on such principles – such as the obligation not to injure (and its counterpoint right of human agents, at least, not to be injured) – have moral grounds for establishing a hierarchy of levels of injury and for punishing the failure to discharge the accepted universal principles on which they are based. This seems to dispose of Baxter's first criticism, in that it identifies a moral basis for regulation to ensure fair discharge of obligations. His second objection, however, is that already noted above, namely that O'Neill's argument fails to define fundamental obligation directly to the non-human world, since the obligations to animals and the environment are indirect or derivative from the impact that damage to them may have on other humans. As a result, humans may release one another from their obligations and effect unlimited damage to the environment. At some stage this could become feasible because humans may conclude that they have ceased to rely on the non-human world in any sense. Baxter concludes that environmental ethics must establish fundamental intrinsic values in the environment in order to provide satisfactory moral grounds for protection of the non-human world from various forms of damage.

11.2.2 A counter to human exceptionalism

This second objection is more difficult to resolve and requires a broader engagement with moral philosophy, in particular, to undermine anthropocentric arguments which centre around an assumption of the uniqueness of humans – the assumption that by virtue of a particular quality (the ability to reason or the existence of the human soul), humans are separated from, and positioned above, the natural world. Peterson (2001, 2) notes that '... environmentally harmful practices, and lifestyles rest on definite, though often implicit, assumptions about human nature ...' of which Western belief systems that define humans as unique are of particular significance. She also identifies several strategies for rejecting this dichotomous view of humans and nature. The most basic of these is the reminder that humans *are* indeed animals and are part of a historical evolutionary process in which there are many connections, branches, terminations and relationships; in this sense, humans are simply one life form *within* nature and not outside it.

> *Evolution teaches ... that we are not only related to other life-forms but also dependent upon them and on the processes and substances of our common environments. The philosophical and ethical implications of this claim are clear: the vision of humans as atomistic and unprecedented lords over a distant 'nature' cannot survive evolutionary science's portrait of the complex, intimate relations among humans, other species, and the physical environment.*
>
> Peterson, 2001, 175

Anthropocentric arguments about ethical value in the environment inevitably begin from the premise of human exceptionalism, and to find real value in the environment it is first necessary to challenge this assumption.

It is, in fact, a particular and relatively recent tradition that separates humans from nature, the origins of which may be traced from the mid-17th century in the development of both Christian theology and the reductionist world view of Enlightenment science. A mediaeval western theology such as that of Thomas Aquinas (1225–1274) was, like modern eastern and indigenous world views (see below), comfortable with a concept of harmonious relations between God and creation, and between human and non-human aspects of the world. However, as the scientific Enlightenment accelerated in the latter half of the 17th century, more entrenched positions emerged, and the not entirely dissimilar theological model of Benedict de Spinoza (1632–1677) was outlawed. Perhaps in part because of their presentation in a kind of Cartesian logic, with propositions, theorems and proofs, Spinoza's views were seen as a direct and radical scientific challenge to Christian theology. In *The Ethics*, Spinoza interpreted the world not as the work of a God who exists independently of it, and who privileges human beings over the rest of (his) creation. Rather, he saw God as immanent in nature and manifest in the laws that govern a natural world of which humans are only a part; for Spinoza, it was wrong to think of humans as creatures outside the causal nexus of nature. The perceived heresies here were innumerable. The deity was 'reduced' to being part of a system of natural law; humans seemed to be deprived of free will; there appeared to be no room for hierarchical, divinely constituted forms of government; and this all laid the foundations for both a *radical* enlightenment in which theology and science were not, necessarily separated, and in which grounds for political and social dissent emerged (Israel, 2001). The truly radical Enlightenment derived as much from this theology as from science, and the theological reponse was, in effect, a reaffirmation of the exceptionalism of both God and humanity, the separation of science from metaphysics and the abandonment of a world view that had survived from the mediaeval period. Ironically, the view of human exceptionalism that this fostered was even more successful than Spinoza in creating the context for an anthropological view of religion, in which deities and religions are themselves interpreted as constructs of the human intellect.

However, although Western intellectual traditions since the Enlightenment have reified humanity, separated it from nature and adopted a reductionist view of the functioning of natural systems, this is not true of indigenous and non-Western knowledges, which 'systematically reject the ontological dualism ... of which the dichotomy between nature and culture is the prototypical instance' (Ingold, 1996). As Peterson (2001) shows, Eastern and American Indian cultures place considerable emphasis on interdependence amongst the many elements of nature, that include humankind, and also display a strong sense of place and create affinity with the environment and a grounded ethic in which there is a duty of care. 'One thing many indigenous cultures emphasize, both practically and ethically, is *relationships* both among persons (of the human and non-human sort) and between persons and the land' (Peterson, 2001, 126). Western feminist philosophy appears to share some of these concerns, representing a critique of human exceptionalism from within the Western tradition. Thus, there are good reasons to question the validity of human exceptionalism and to examine the ethical significance of the duty (obligation) of care (Tronto, 1993) for an environment which forms a web of relationships within which humans and their natural environment co-exist. Such an examination may reduce the need to call upon anthropocentric starting points in the search for ethical value in the environment, and recalls Aldo Leopold's (1949, 239–40) 'land ethic', which '... enlarges the boundary of the community to include soils, waters, plants, and animals, or collectively: the land ... [and] ... the role of *Homo sapiens* [changes] from conqueror of the land-community to plain member and citizen of it'.

Various social constructionist views of nature, which treat the environment as a product of human action and thought, appear initially to reinforce human exceptionalism. However, they can be shown, by contrast, actually to undermine the separation of humans and nature. Constructionism may, at one level, simply mean that 'nature' is the result of thousands of years of human impact, a truism which hides the distinction between product and process. That is, the visible appearance of the landscape and environment occupied by humans may be a *product* of their cumulative impacts, but unobservable *processes* continue to operate in that environment much as they have since pre-history, except, perhaps, in terms of their rates of action. In other words, human impact is an effect in degree, but not in kind; the greenhouse effect is not a human impact, but its current intensity may be (although evidence from ice core records of atmospheric chemistry and late Quaternary climatic change does not entirely support even this conclusion – Anklin et al., 1997). Thus although human impact seems to imply a humanity external to nature, it transpires that humans are simply in large measure a component of that nature, a factor within biogeochemical cycles. This, of course, is the kind of argument that underpins Lovelock's Gaia hypothesis (1979).

There is also a more extreme form of constructionism, in which there is no 'nature' beyond our mental constructs, with different conceptions being equally valid, and the best scientific knowledge presenting but one model of nature that cannot be privileged above that of traditional or indigenous knowledges. Again, this view appears to support a human exceptionalism, since usually only humans are considered able to construct a view of their world, through their intellect and their capacity for reason and communication. This extreme position needs the careful analysis undertaken by Hacking (1999) who, in a wide range of contexts, shows that when 'things' are said to be socially constructed, it is only in special circumstances that this is true, and that there are frequently very good grounds for believing those 'things' also to be very 'real', and outside their human constructs. An obvious example is the natural disaster, a realization of an environmental hazard. This is only a hazard, or disaster, because it is perceived as such by the human beings affected by it. However, it is also a real, natural event which is independent of humanity, and which can happen without people being present to perceive or 'construct' it. A debris fan is created by infrequent debris-flow events, but these processes are quite normal in mountainous and semi-desert environments, and indeed are the processes to which such fans are adjusted (Beatty, 1974). These events are 'catastrophic', however, if humans in the landscape are affected by them. For Hacking, the question is therefore: the social construction of . . . what?

The answer is that the environment is real, and 'out there', but that how humans perceive and construct it is internal and culturally specific. While different constructions deserve respect because they have different cultural contexts and purposes, some are privileged in particular circumstances; human beings would not have walked on the moon had they (all) continued to believe that this celestial object was a calabash (to borrow an example from Richard Dawkins, 1994). But since a diverse range of constructions of nature, landscape and environment at least deserve respect, there is no particular reason to view the human capacity to construct as exceptional. Thus, if it can be shown that non-human animals also construct mental images of the nature in which they live, the constructionist perspective actually provides grounds for dissolving human exceptionalism. There is, in fact, plentiful evidence for social behaviour and communication within animal populations, and for behaviour that indicates awareness of the resource base of their environment – indeed, evidence for forms of reasoning. As a result, there are no ethical grounds for privileging human constructions of nature relative to those of other animals, or indeed, for privileging humans. Since animals are irrevocably linked to inanimate nature – to plants, ecosystems and landscapes – there are good reasons to accept that there is a continuum including humans, animals, plants and inanimate nature, and no grounds for defining a sharp boundary between humans and the rest of nature. This is not a continuum on a univariate (or, indeed, multivariate) quantitative scale, susceptible to group identification by discriminatory analysis; it is multi-dimensional, qualitative and indivisible. Humans may construct aspects of nature, but the opposite is also true; and humanity is but one of the many faces of nature. Identification of this continuum opens the possibility for establishing those attributes of this composite nature that define common ethical value, and gives the environment ethical standing without recourse to anthropocentric arguments, since such arguments rely on a dubious claim of exceptional status for humans with respect to nature.

11.2.3 The value of relatedness

In the above discussion, two key themes which emerge are those of 'obligations' and 'relatedness', and these provide good grounds for the determination of real environmental ethical values. In essence, this involves positing a human ecology in which these values are identifiable in the networks of processes which underpin the relatedness between and amongst human and natural environmental attributes. The ecosystem model can thus be extended to include both human and human–environment inter-relationships in a form of human ecology (Chapman, 1977). In this environmental ethic, it is nature that leads and humans that follow, simply because humans are a part of nature and the charge of anthropocentrism is laid to rest. An important property of natural ecosystems is their many evolved symbiotic relationships, which are emergent properties that transcend quantifiable considerations of energetics. These may be characterized in some cases as evolved obligations, the severance of which has severe implications for ecosystem stability. Critical attributes of the ecological framework are thus the networks of attributes, mechanisms and processes that provide the structured linkages between individuals, groups and their environments. It is then, ultimately, these structures which may be suggested to provide the source of the 'real' values residing in the environment, and these structures are what provide the *gestalt* properties of the whole system of nature (including humanity). Thus, the argument leads to some level of identification of what it may be that is the source of real ethical value in the environment. Ontologically, it is of course difficult to justify giving privilege to unobservable structural relationships relative to observable entities (such as certain trees and animals), since both are essential to the whole, and to do so merely replaces one partial view with a different one. However, when single observable entities are injured directly, the whole may be more recoverable than when indirect injury undermines the entire system function. In this sense, *indirect* effects on classes of observable entity through the structures that link them may be more systemic and lasting, and thus more critical to the system as a whole.

The ecosystem concept has several advantages as a basis for geographical enquiry (Stoddart, 1967). It does permit the bringing together of humans, animals, plants and the environment in a single framework. The structured nature of an ecosystem readily allows rational investigation; and ecosystems function as a result of matter and energy transfers, which may also be monitored within the framework of an ecosystem. However, an additional attribute is that ecosystems involve mutual dependencies, in which components may almost be said to display rights and obligations towards one another as a consequence of their interdependent evolution and adaptation. Indeed, it is common to refer to certain species as having 'obligate' status. For example, the genus *Salicaceae* consists of obligate riparian species, adapted to the moist conditions adjacent to rivers. Functional analyses in terms of energy flows, trophic levels, and food webs underpin many important inter-relationships within ecosystems, but do not capture the full details of these adaptations, mutualisms, co-existences, co-evolutions and dependencies. Plant associations that create familiar assemblages (communities) arise because of the exploitation by one species of the niches created by others; this may be termed the 'social life of plants'. A significant proportion of the angiosperms have species-specific parasitic relationships with other plants (Press and Graves, 1995). Faunal species also find opportunities for 'peaceful coexistence' (Collinvaux, 1980), adapting their food preferences to avoid competition by foraging in spatially separate locations. Flora and fauna have co-evolved to facilitate one another's reproduction and feeding. For example, the bills of some specialized hummingbirds match the flowers of the species which they are adapted to pollinate (Snow, 1981; Figure 11.1). Certain acacia seeds only germinate when they have passed through the gut of an ungulate, and the wild tomato *Lycopersicon cheesmanii* Riley on the Galapagos Islands only regenerates when its seeds are excreted by the tortoise which feeds on it (Rick and Bowman, 1961). These relationships arise where seeds have hard surfaces that must be physically abraded during digestion or where germination-inhibitors are chemically leached in the gut. The seed is then deposited in a moist, nutrient-rich medium which encourages germination and establishment, and the mobility of the temporary host ensures dispersal. A beautiful illustration of natural obligation involves the Mauritian tree *Sideroxylon sessiliflorum,* which regenerates only rarely. The fruits of this tree were ingested and dispersed by the dodo (which became extinct in 1681), and the seed germinated when excreted. The dodo had an obligate relationship with this species, and this relationship has been destroyed along with the dodo. However, some limited success has been achieved by feeding the fruits to turkeys, in the absence of alternative mega-pigeons to the dodo. There are innumerable examples of these relationships amongst species in ecosystems, which betoken high levels of self-organization, and emergent states of mutual dependence which cannot simply be accounted for by the energy relationships between species.

It is a relatively small additional step to extend the ecosystem concept to include human beings. There have, in fact, been many arguments in geographical methodology in support of the case that human beings and the natural world are irrevocably

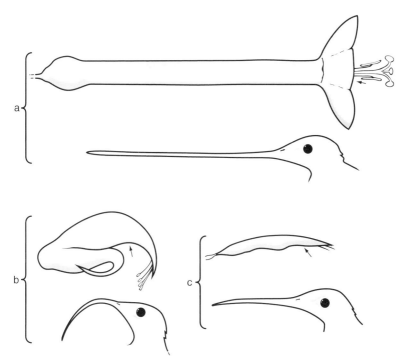

Figure 11.1 *The adaptation of the shape of the bills of some specialized hummingbirds to the flowers that they pollinate. The species are (a) the swordbill and the passion vine; (b) the sicklebill and Heliconia sp; and (c) the velvet-breast and Castilleja fissifolia. Arrows indicate the point of entrance of the bill (From Snow, 1981.)*

interdependent (Barrows, 1923; Morgan and Moss, 1965; and Chapman, 1977), and some of these have invoked a form of ecological analysis which has been termed 'human ecology'. Eyre and Jones (1966), for example, made a case for human ecology, but interestingly do not invoke the ecosystem concept to any significant degree in doing so. For them, human ecology is simply concerned with 'interaction between human activity and natural circumstances' (Eyre and Jones, 1966, 6), and is reflected in spatial associations of attributes rather than system structures and the mechanisms underpinning interaction. A more compelling use of the ecosystem concept in the study of relationships amongst human beings and the natural world would, however, focus on these processes of interaction and mutual adaptation. The key phenomena here are the networks of inter-relationships, which in turn lead to the centrality of forms of reciprocity, so the argument returns to the direct and indirect obligations and rights of humans for one another and for nature. The natural ecosystems then involve a parallel form of reciprocity, because an intrinsic part of an ecosystem is a form of obligation–rights relationship amongst non-rational, but active, evolving and adapting entities. Thus if there is an obligation not to harm part of an ecosystem in order to sustain an inter-relationship, the links between that structure and others will mean that other elements of the system are also protected or conserved. This conceptualization of nature and humans in a unified and holistic structure may, of course, appear to be most immediately relevant to those traditional societies which are heavily dependent on locally available natural resources, and which retain the kinds of integrated, indigenous ethical frameworks discussed above. However, it has become increasingly relevant also to industrial and Western economies, and at global rather than local scales, as the effects of trans-boundary pollution and climatic changes linked to fossil fuel use have become more widely understood. This indicates that the ecosystem model does not have to imply a static view of the world, with fixed and immutable structural relationships. Quaternary climatic changes have shown that ecosystems undergo continual adaptations to change. However, present rates of climatic and human-induced change occur at rates which tax the resilience of the structures that maintain system integity, and threaten irreversible degradation both of the environment and of social well-being.

This, therefore, seems to be where the 'real' ethical value of the environment lies, a value which O'Neill (1997) has concluded is elusive and only given apparent substance through rhetorical claims. The values in the environment, or in nature, are found not in tangible entities or observable events, but in unobservable structures and mechanisms; in the phenomena which underlie and account for the relationships amongst and between humans and nature. Indeed, the discovery of real value in the environment seems to require a realist ontology (Bhaskar, 1989) to focus on the unobservables that create and maintain the integration of an ecosystem. It may not be possible to claim ontological priority for these unobservables, since entities, events, structures and mechanisms are all part of the whole which is the integrated human ecology. However, to assess the consequence of human actions within this ecology, it is necessary to understand the unobservables, since these are the active elements of the whole structure – the source of its 'becoming' rather than the components of its 'being' – and the elements whose disturbance may have far-reaching effects. Tacitly, this is recognized in a whole range of human institutions for environmental management and human development. The precautionary principle is a policy tool which recognizes the risks inherent in intervention in an environment whose structural relationships are imperfectly understood. Contemporary practices in development involve community-based, participatory approaches which can draw on indigenous environmental knowledges, help to maintain local social capital and preserve the entitlements, institutions and evolved practices that help to sustain both environment and society (Nelson and Wright, 1995). But when local common property resources are appropriated for external markets, the potential exists for social collapse and environmental degradation (Hviding and Baines, 1994), as the networks of structural linkages amongst people and their environment are broken. The following case study illustrates such a circumstance.

11.3 Human ecology in the Awash Basin

River basin management provides a rich source of material to exemplify structural integration of human and environmental processes. Its guiding principle since the establishment of the Tennessee Valley Authority in 1933 has been the joint management of land and water resources for the sustainable benefit of society. The rationale for this was clear when soil conservation was urgent, water management was a priority for economic development and the success of one was reliant on that of the other. This integration, reiterated by Newson (1992), underpins many studies of water resource use in drainage basins in developing countries. It is less clear, however, that the needs of all human population groups reliant on drainage basin land and water resources are treated with a comparable level of integration, or that they are even all treated ethically. Indeed, an evident lack of ethical treatment of humans underpins Roy's polemic about the Sardar Sarovar dam project on the Narmada River (1999). The case study discussed below also draws attention to the loss of an essential and overriding ethical dimension in resource development when the human ecology of a catchment is ignored; that is, when the mutual dependencies of human groups are disrupted, rights and obligations are destroyed and injury follows. This consequence can arise when planning adopts a reductionist approach with separate specialist investigations being combined, but the networks of interdependent environmental and social processes not being fully represented. The aim of the case study is, therefore, to show that the human ecology of a drainage basin must be understood before development intervention, and that failure to achieve this may have demonstrably unethical results.

The case study is of water resource development in the Awash valley in Ethiopia. The Awash (11.2) is a major river which heads in the Ethiopian highlands near Addis Ababa, and descends into the East African Rift Valley, to flow north-eastwards until its dwindling waters evaporate in the internal drainage of Lake Abhé. Exploitation of the water resource of this river and its tributaries has occurred since 1950, together with associated development of irrigated agriculture (mainly of sugar and cotton) on the Awash floodplain (Figure 11.3). This began with creation of the Wonji sugar estate, construction of the Koka Dam in 1960, and establishment of the Awash Valley Authority in 1962. By 1973, 77% of the irrigable area of the upper valley was developed, 43% in the lower plains and 14% in the middle valley (Kloos, 1982).

11.3.1 A human ecology and its destruction

Prior to 1960, there was little evidence of land degradation in the Awash valley, and the dominant mode of production in the valley bottom involved transhumant pastoralism by essentially settled populations (*not* nomadism). Several groups exploited seasonally inundated floodplain areas as communal dry season grazing and for flood recession cultivation. The Jile herded cattle and some sheep and goats, and grew maize and beans on the floodplain at Wonji; the Arsi used the Nura Era floodplain; the Kereyu had dry season villages in the Abadir-Metahara region (Ayalew, 2001); and the dominant Afar group herded cattle on the floodplain areas downstream from Awash Station. All these pastoralists practised seasonal transhumant migration. In the wet season (June–September), they moved away from the flooded areas, which were mosquito-

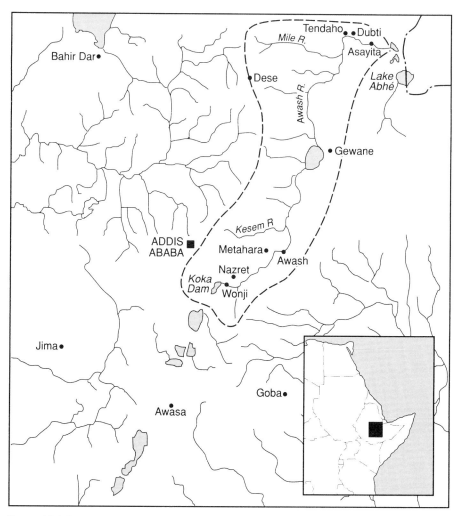

Figure 11.2 *The Awash River and its catchment, Ethiopia.*

infected, over distances of up to 200km. The Jile moved into the local foothills of the southern escarpment; some Arsi moved to the Somali plateau; the Kereyu to the wet season pastures on the savanna of the Kesem-Awash watershed; and the Afar moved both locally to higher savanna and over longer distances to the central highlands. In the October–May dry season they returned to the floodplain areas where nutritious, semi-aquatic grasses (*Echinochloa* and *Voatia cuspidata*) supported higher stock densities than the fire-resistant grasses of the savanna outside the floodplain. The riparian woodland provided browse for camels and goats, and its undergrowth extra forage for sheep, cattle and donkeys.

The pastoralists traded their livestock and livestock products for highland grain in a series of markets along the foothills of the north-eastern escarpment (McCann, 1987). A

Figure 11.3 *The development of commercial irrigated agriculture in the lower Awash Valley.*

contractual system existed in which highland agriculturalists bought cattle from the Afar, who then acted as herdsmen in return for the milk and some calves. Highlanders gained the oxen needed to pull the heavy, two-oxen *maresha* plough as a result of this trading system (Maknun Gamaledinn, 1987). Since approximately 65% of peasant households in Shoa and Wello then had either none or only one ox, systems for purchase, hire or loan of the draught animals needed to pull the plough were critical. Pastoralists also relied on grain 'bought' from highland farmers to supplement their nutritional requirements. Crucial trading linkages therefore cut across the ecological zones defined principally by altitude, with the need to gain access to resources supplied by the other zones providing the dynamic for well-established social and economic relationships. Here, therefore, is an excellent example of a human ecology, involving symbiotic relationships of the kind observed in ecosystems, taking advantage of the different resource bases offered by the environments of different ecological zones, and reflected in counterpart rights and obligations.

This highly adapted system of exploiting and conserving the resources provided by an otherwise uncertain and sometimes marginal environment was badly disrupted by the development of commercial irrigated agriculture in precisely the floodplain areas which were the mainstay of the pastoralists' economy (Girma Kebbede and Jacob, 1987). Even where floodplain lands were not converted, the grazing potential declined because control of the flow of the Awash by the Koka Dam reduced the proportion of the total annual flow discharged in July to September from 84% (1943–58) to 35% (1962–72), with an attendant reduction of the flooded area (Kloos, 1982). Deterioration of the riparian woodland occurred as the water table declined and the floodplains became increasingly arid. In some areas, migration routes to wet season pastures were blocked (at Nura Era, Melka Sadi-Amibara). Woodland was cut to provide timber for building on the new farms, and new road access accelerated the use of woodland to produce charcoal for the more accessible urban market. Pastoralists had no choice but to move to ecologically more fragile areas, where land degradation occurred rapidly because previously 'safe' stocking levels resulted in overgrazing, and where they became vulnerable to the effects of drought. Bondestam (1974) noted that '... this relative overpopulation on the remaining land is a recent phenomenon, caused by the outside-initiated and controlled commercial development. Any other explanation is false'. Land degradation was accentuated in the foothills of the north-eastern escarpment because westward-migrating pastoralists *and* agriculturalists moving down from the plateau both settled here, giving rise to the highest population densities in the Awash Valley (100–150 per km^2).

An additional consequence of the development of commercial agriculture and its effects on pastoralism was a change in the characteristics of the rural market system. Changes that occurred after 1960 disrupted the patterns of exchange noted above, and this was accentuated by the growth of an urban market for grain, which began to dominate the traditional trade between agriculturalists and pastoralists and inhibited transfer of surplus production to rural deficit areas (Horowitz and Little, 1987). Commercial agriculture tended to substitute export cash crops for food production and inhibited the distribution of food within the rural populations which were most vulnerable to the effects of drought. The result was an increasingly severe impact of drought; estimates suggested that 25–30% of the Afar died in the first catastrophic 1972–73 drought (Maknun Gamaledinn, 1987), and neither this nor the 1984 famine were disasters entirely attributable to the vicissitudes of climate. As Kloos (1982, 40) stated:

> ... man-made famine is a recent phenomenon in the Awash Valley. The available evidence suggests that the 1972–3 famine was not caused entirely or primarily by failure of the rains or by deliberate overstocking by local pastoralists, but rather by dam construction, large-scale irrigation development that relied almost exclusively on migrant labourers from the highlands, and ensuing loss of grazing and water resources, and environmental degradation.

To this may be added severe disruption of both the system of exchange and resource transfer in the rural economy, and the destruction of the negotiated rights and obligations that sustained this economy. Here, therefore, is evidence of an imposed development path whose consequences were such that it may be regarded as fundamentally unethical, because, in the most damaging manner possible, it caused injury to both the human and non-human (environmental) elements of an intricately constructed (human) ecology.

11.3.2 Environmental degradation and the space–time dynamic of resource use

The changes outlined above are by no means unique. Similar effects have arisen from irrigation development of the former dry-season grazing reserves of pastoralists in Sudan, Niger, Kenya and Tanzania. Thus, there are several precedents for the interpretation of events in the Awash Valley. Indeed, environmental degradation, of both soil and ecology, is often blamed on poor agricultural practices by indigenous cultivators, or on overgrazing by pastoralists. However, indigenous peoples have often been marginalized, both economically and spatially, by the development of more capital-intensive agriculture, being forced to settle less productive, erodible land as a result of both external and internal colonial processes, in which ruling ethnic or political groups impose economic and development policies on peripheral and subordinate groups. This marginalization has often been a major reason for environmental degradation (Figure 11.4) and, in extreme cases such as that described above, of famine. Severe soil erosion in several parts of southern Africa has been attributed to marginalization of indigenous cultivators onto erodible soils (Darkoh, 1987). A particular twist is given to the story when the degradation resulting from colonial expropriation of land then leads to conservation institutions (Anderson, 1984). Even in the USA, Graf (1986) has been able to show how the politics of water resource development in the Colorado basin in the early 20th century led to the

Figure 11.4 *Evidence of land degradation in the upper Awash catchment in the late 1980s; a small village reservoir impounded by an earth dam, completely filled with sediment as a result of soil erosion in its headwaters.*

establishment of the Soil Conservation Service as a device to impose stock control on the Navajo Nation, in spite of emerging scientific evidence that subtle climate changes in sensitive climatic regions were more likely explanations for gully erosion. Thus land degradation is invariably the consequence of a complex mix of environmental and socio-political changes, usually involving destruction of relatively robust human ecologies by some form of unequal power relationship of doubtful ethical standing.

The systems diagrams used by Blaikie (1985) to illustrate the multi-disciplinary nature of the analysis of land degradation demonstrate well the complex interdependence of social and environmental processes involved. An example of such a diagram is that by Franke and Chasin (1981), representing the human ecology of peanut production in Niger, and its negative effects on food security and land management (Figure 11.5). Interestingly, Figure 11.5 was first published in the journal *The Ecologist*. Systems diagrams of this kind tend to imply a set of functional relationships, and although showing the intricately structured inter-relationship of natural and social processes, tend to suppress evidence of the spatio-temporal nature of the functioning of the structure as a whole. However, if it is recognized that each of the variables represented as a box in the diagram has a spatial distribution – in other words, each 'box' holds a map – then it is possible to understand how the structure works, and what is required of *inter*-disciplinarity in its analysis. There are different issues of concern, such as resource appraisal, resource allocation and resource use. These must all be continually revisited in the representation of the history of a structure which represents a resource management question, as the spatial distribution of productive resource is altered both by its prior use and abuse, and by changing demands and technologies, and as external factors also change over time and space (including global commodity prices, regional climate and local politics). A truly inter-disciplinary approach arises only when the spatio-temporal dynamic of a complex system is analysed, since then it is impossible to hide behind the technical language of a partial investigation and to dismiss supposedly peripheral issues in a vernacular language. This is therefore a refutation of Johnston's (1983) critique of inter-disciplinarity quoted at the beginning of this essay, and exposes his argument as one which assumes a descriptive geography of a static landscape, rather than the spatio-temporally dynamic landscape with which it is analytically concerned.

11.3.3 Development economics and ecological destruction

Part of the reason for the recurrent imposition of sedentarization and privatization in formerly transhumant/nomadic systems and communal pastoral areas is the widely held view that such a change is 'progressive', replacing a low-productivity livestock sector which has few possibilities for increased production. This is reinforced by the assumption (often discredited; Graf, 1986) that nomadic pastoralists cause overgrazing, and by the tendency for administrators to be drawn from ethnic groups with roots in sedentary agriculture. However, Horowitz and Little (1987) have actually questioned the economic assumptions underlying this view. First, while conceding that the productivity of

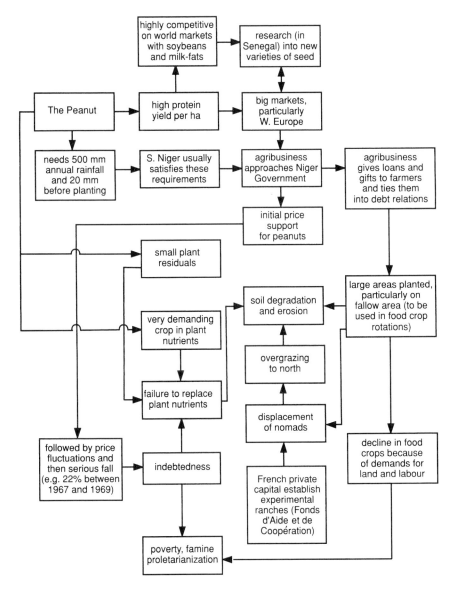

Figure 11.5 *The multi-disciplinary nature of the human ecology of peanut production in Niger, and its negative effects on food security and land management (after Franke and Chasin, 1981).*

pastoralism is low, they present evidence that traditional peasant pastoralism is substantially more productive than commercial ranching. In southern Ethiopia, Borana pastoralism produced four times the protein and six times the food energy per hectare of Australian ranches, and the cost (labour etc.) per unit of protein production is about a tenth of that in Kenyan and Australian commercial ranches. (Ranching usually measures its productivity per head of animal, which is a more favourable statistic but arguably a less useful basis for comparison with other systems of production.) Economic criteria for judgement about the value of a mode of production are thus inherently contestable, and arguably less stable than ethical judgement about the production system in its environmental and social context. Second, estimates of livestock-carrying capacity are notoriously difficult. In Ethiopia, it is estimated that one hectare will support one TLU (Tropical Livestock Unit) where rainfall exceeds 1000mm, while 22 hectares are required where rainfall is about 200mm. This, however, ignores the role of exogenous river flow, and the complicating effect of the extreme year-to-year variability that can occur in rainfall in drier regions. Traditional systems provide a resilient and adaptable method of exploiting variable forage resources, maximizing production in wet years rather than remaining at low stock densities which are effectively understocked. Although the lag before stock levels decline in dry years may cause some land degradation, there is evidence that persistent understocking may reduce the quality of pasturage as non-palatable thorny bushes invade and tsetse fly infection increases because the canopy becomes more extensive. This is another case where the impacts of change can only be understood in holistic ecosystem terms, and through analysis which is sensitive to regional and local social, cultural, economic and environmental contexts.

In the late 1980s, a further feasibility study was conducted to establish the potential for additional development of the Awash basin's water resources. This involved a team of experts: engineers (in hydraulics, irrigation and power generation), geographers, remote sensing experts, geologists, agronomists, economists, sociologists and anthropologists, ecologists and hydrologists. The study included an analysis of the spatial pattern of soil erosion in the Awash valley, to assess the areas most likely to suffer land degradation through loss of ground cover in the near future, to explore the potential for implementation of soil conservation measures, and to evaluate potential reservoir sites in relation to the sedimentation rates likely to be experienced because of continuing soil erosion (Griffiths and Richards, 1990). Thus there was explicit concern for the integrated analysis of land and water. However, it is not so evident that there was sufficient examination of the human ecology. Contributors to the study (including the author) were generally required to act as narrowly defined, reductionist specialists rather than to examine in a holistic sense the systematic structural interdependencies amongst human agents and catchment environment. This issue must inevitably be addressed in future management of water resources, in the Awash valley as elsewhere, because the droughts and famines that afflict Ethiopia continue to have the capacity to injure, and to receive international media attention, and the consequences of a lack of ethical human ecology in the planning are rapidly exposed. In view of the needs to sustain indigenous subsistence food production in Ethiopia (of both crop and livestock products in traditionally integrated sectors), to maintain the viability of rural markets which maintain the links between those sectors and to reduce the susceptibility of the peasant population to drought and famine, radical management and development strategies are required. Development which emphasizes cash crops for export and centralized marketing is unlikely to achieve these objectives. The interests of pastoralists merit closer attention, both because their traditional production system was well adapted, and because imposition of an alien and disruptive system may generate greater opposition than the co-operative introduction of sustainable livestock schemes. Appropriate land tenure arrangements for pastoralists, appropriate scales of livestock production units for development (i.e. based on household, clan or tribal groups), appropriate support services (e.g. veterinary) and appropriate marketing methods, which link the small-scale production of pastoralists and agriculturalists, all need to be considered. These would be consistent with contemporary emphases on community-scale, participatory development processes, in which locally negotiated, democratically determined, 'bottom-up' outcomes are sought that are owned by those who are directly affected, and who are often best qualified to determine how to preserve, but develop appropriately, a human ecology. This is consistent with Smith's (2000, 211) argument that 'community … [is] … an essential ingredient in moral motivation'.

11.4 Conclusions

Several conclusions may be drawn from both the general ideas and the specific example discussed above. The first is that ethical grounds do indeed exist for an integrated, geographical human ecology, in which concerns with humanity and environment are appropriately balanced, and the mechanisms sustaining obligate relationships are the subject of analytical enquiry. Such a focus allows for the kind of emancipatory realism envisaged by Bhaskar, in which the unobservable structures and mechanisms underlying an issue must be revealed, in order both to prevent injury through mis-management, and to ensure that social transformation can be rooted, ethical and freed from '… the dichotomous opposition between nature and society' (Bhaskar, 1989, 6). An integrated geography should be an ethical geography which values the structural relationships between humanity and nature; and an ethical geography must be integrated.

The second conclusion is that the interplay between society and environment is locally contingent, so that one additional role for an integrated geography is to understand the place-sensitive variations of these structural relationships. There can be no simple, common panacea for a particular problem, a conclusion nicely illustrated by the different experiences of Machakos and Embu Districts in Kenya. Tiffen et al. (1994) show that in Machakos, environmental degradation evident in the 1930s had been reversed by the 1990s, in spite of continued population growth, through a combination of practices which conserved both tree cover and soils, and by improved physical access to, and information about, market opportunities. In Embu, Joekes et al. (1994) suggest that the social structures, particularly those related to women's entitlement to land ownership, have conspired with the environmental stress associated with drought to cause environmental deterioration. The need to examine how particular structures and their contexts vary from place to place implies that an appropriate scale exists for integrated geographical analysis. This scale is that at which a community can be identified which is capable of displaying moral motivation and owning the practice of environmental valuation; the scale espoused, for example, by Peterson (2001) and Smith (2000).

However, a third conclusion must be that the scale of 'community' varies from problem to problem. Global issues now have considerable currency, as the human community is seen to be capable of disturbing the interdependent physical and biological systems of global biogeochemical cycles, altering the intensity of the greenhouse effect and the climate. Lovelock's (1979) Gaia hypothesis provides a framework within which to evaluate such global-scale human influences on biogeochemical cycles, stores and fluxes, and the mechanisms embodied in the hypothesis account for the evolution of the earth–atmosphere–ocean system at scales that dwarf human impacts (van Andel, 1994). One reaction to Gaia invoked metaphysical implications about the earth, almost as a global version of an eastern or indigenous people's world view. However, it is evident that the global community needs much stronger institutions than a new-age spiritualism in order to regulate trans-boundary environmental problems and other relationships of humanity and nature in a globalized world. An integrated geography would have to confront the ethics of differential nation-state responses to the international institutions and global governance demanded by the climatic and oceanographic changes which follow from the fossil fuel economy. At the same time, it has to recognize that '... the roots of global environmental problems are local, and their solution requires linking local with global perspectives' (Dasgupta et al., 2000, 344), and that institutional reform is necessary at several scales to create incentives for individuals and groups to operate for the common good.

An additional conclusion is that ethical considerations must underpin policies applied both to environmental management and to human development. This demands policy based on understanding of structural relationships, not simply on observable states and events. In the field of environmental policy there are already examples of this. The precautionary principle, the 'polluter pays' principle, and attempts to internalize externalities are all examples, as is the gradual development of international accords (such as the Montreal Protocol). However, the challenge for these policies is to make them structurally relevant, but place-sensitive, as argued by Trudgill and Richards (1997). One way of helping to achieve this is by making elements of the policy adaptable following community involvement, drawing on local expertise and knowledge in ways that may reduce the erroneous application of blanket responses to environmental management needs. This has been attempted with some success, for example in flood management practices in Australia and Europe which maximise ownership of their planning and implementation by the relevant tier of local government (Handmer, 1996), and in schemes that integrate unofficial warning systems with official ones (Parker and Handmer, 1998). These moves represent an environmental policy equivalent of participatory, community-led development strategies, and are rapidly being adopted by government institutions such as the bodies for environmental regulation and management.

To conclude, this essay mixes professional interests and personal experience. The former concern the search to justify integration in geography from a theoretical perspective, and to explore the roles that ethics, realism and holistic analysis can play in helping to construct this integration. The latter arises from practical application of this form of knowledge, particularly in the context of experience as an environmental consultant. Together, these enable a renewed and more fundamental case to be made for a less reductionist, more holistic geography, which is a form of 'human ecology' based on analysis of the networks of mechanisms (both environmental and social) that create structured forms of inter-relationships between environmental and social entities. The empirical case study demonstrates that an ethical development process must take note of these networks, must avoid the abrupt severing of critical connections which maintain stable relationships (although equally, must not become paralysed by the assumption that change cannot occur), and must ensure the kinds of participatory involvement that can prevent this. The arguments constructed in this paper therefore attempt to provide the basis for an applicable human ecology which is at the same time a form of moral geography.

Acknowledgements

I am grateful to Nick Clifford, David Livingstone, David Smith and Liz Watson for extremely thought-provoking and helpful exchanges which may well lead to more in this vein, but the usual defence of such discussants applies; errors, logical inconsistencies and weaknesses in the argument are all my fault.

References

Anderson, D. 1984: 'Depression, dust bowl, demography, and drought: the colonial state and soil conservation in East Africa during the 1930s'. *African Affairs* 83, 321–43.

Anklin, M., Schwander, J., Stauffer, B., Tschumi, J., Fuchs, A., Barnola, J.M. and Raynaud, D. 1997: 'CO_2 record between 40 and 8 kyr BP from the Greenland Ice Core Project ice core'. *Journal of Geophysical Research* 102, 26539–26545.

Ayalew Gebre 2001: *Pastoralism under Pressure: Land Alienation and Pastoral Transformation Among the Karrayu of Eastern Ethiopia, 1941 to the present.* Shaker Publishing BV, Maastricht.

Barrows, H.H. 1923: 'Geography as human ecology'. *Annals of the Association of American Geographers* 13, 1–14.

Baxter, B. 1999: 'Environmental ethics – values or obligations? A reply to O'Neill'. *Environmental Values* 8, 107–112.

Beatty, C.B. 1974: 'Debris flows, alluvial fans, and a revitalised catastrophism'. *Zeitschrift für Geomorphologie, Supplement Band* 21, 39–51.

Bhaskar, R. 1989: *Reclaiming Reality: A Critical Introduction to Contemporary Philosophy.* Verso Press, London.

Blackburn, S. 2001: *Being Good.* Oxford University Press, Oxford.

Blaikie, P. 1985: *The Political Economy of Soil Erosion in Developing Countries.* Longman, Harlow.

Bondestam, L. 1974: 'People and capitalism in the North-Eastern lowlands of Ethiopia'. *Journal of Modern African Studies* 12, 423–39.

Chapman, G.P. 1977: *Human and Environmental Systems: A Geographical Appraisal.* Academic Press, London.

Collinvaux, P. 1980: *Why Big Fierce Animals Are Rare.* George Allen & Unwin, London.

Darkoh, M.B.K. 1987: 'Socio-economic and institutional factors behind desertification in southern Africa'. *Area* 19, 25–33.

Dasgupta, P., Levin, S. and Lubchenco, J. 2000: 'Economic pathways to ecological sustainability'. *BioScience* 50, 339–345.

Dawkins, R. 1994: 'The moon is *not* a calabash'. *Times Higher Education Supplement*, Sept. 30, 1994, 17.

Dear, M. 1988: 'The postmodern challenge: reconstructing human geography'. *Transactions of the Institute of British Geographers* 13, 262–274.

Eyre, S.R. and Jones, G.R.J. (eds.) 1966: *Geography as Human Ecology: Methodology by Example.* Edward Arnold, London.

Franke, R. and Chasin, B.H. 1981: 'Peasants, peanuts, profits and pastoralists'. *The Ecologist* 11, 156–168.

Girma Kebbede and Jacob, M.J. 1987: 'Drought, famine and the political economy of environmental degradation in Ethiopia'. *Geography* 7, 65–70.

Graf, W.L. 1986: 'Fluvial erosion and federal public policy in the Navajo basin'. *Physical Geography* 7, 97–115.

Griffiths, J. and Richards, K.S. 1990: 'The application of a simple Geographical Information System to soil erosion and soil conservation studies'. *Land Degradation and Rehabilitation* 1, 241–262.

Hacking, I. 1999: *The Social Construction of What?* Harvard University Press, Cambridge, Mass.

Haggett, P. 1967: *Locational Analysis in Human Geography.* Edward Arnold, London.

Handmer, J. 1996: 'Policy design and local attributes for flood hazard management'. *Journal of Contingencies and Crisis Management* 4, 189–197.

Horowitz, M.M. and Little, P.D. 1987: 'African pastoralism and poverty: some implications for drought and famine'. In M.H. Glantz (ed.), *Drought and Hunger in Africa*, Cambridge University Press, Cambridge, 59–82.

Hviding, E. and Baines, G.B.K. 1994: 'Community-based fisheries management, tradition and the challenges of development in Marovo, Solomon Islands'. In Dharam Ghai (ed.), *Development and Environment: Sustaining People and Nature*, Blackwell, Oxford, 13–39.

Ingold, T. 1996: 'Hunting and gathering as ways of perceiving the environment'. In R. Ellen and K. Fukui (eds.), *Redefining Nature: Ecology, Culture and Domestication*. Berg, Oxford, 117–155.

Israel, J 2001: *Radical Enlightenment: Philosophy and the Making of Modernity, 1650–1750.* Oxford University Press, Oxford.

Joekes, S., with Heyzer, N., Oniang'o, R. and Salles, V. 1994: 'Gender, environment and population'. In Dharam Ghai (ed.), *Development and Environment: Sustaining People and Nature*, Blackwell, Oxford, 137–165.

Johnston, R.J. 1983: 'Resource analysis, resource management and the integration of physical and human geography'. *Progress in Physical Geography* 7, 127–146.

Johnston, R.J. 1986: 'Four fixations and the quest for unity in geography'. *Transactions, Institute of British Geographers, New Series* 11, 449–453.

Kloos, H. 1982: 'Development, drought and famine in the Awash Valley of Ethiopia', *African Studies Review* 25, 21–48.

Leopold, A. 1949: *A Sand County Almanac, and Sketches Here and There*. Oxford University Press, New York.

Lovelock, J.E. 1979: *Gaia: A New Look at Life on Earth*. Oxford University Press, Oxford.

Maknun Gamaledinn 1987: 'State policy and famine in the Awash Valley of Ethiopia: the lessons for conservation'. In D. Anderson and R. Grove (eds.), *Conservation in Africa: People, Politics and Practice*, Cambridge University Press, Cambridge, 327–44.

Massey, D.1999: 'Space-time, "science" and the relationship between physical geography and human geography'. *Transactions of the Institute of British Geographers* 24, 261–276.

McCann, J. 1987: 'The social impact of drought in Ethiopia: oxen, households, and some implications for rehabilitation'. In M.H. Glantz (ed.), *Drought and Hunger in Africa* Cambridge University Press, Cambridge, 245–67.

Morgan, W.B. and Moss, R.P. 1965: 'Geography and ecology: the concept of the community and its relation to environment'. *Annals of the Association of American Geographers* 55, 339–350.

Nelson, N. and Wright, S. (eds.) 1995: *Power and Participatory Development: Theory and Practice*. IT Press, London.

Newson, M. 1992: *Land, Water and Development: River Basin Systems and their Sustainable Management*. Routledge, London.

O'Connor, M. 2000: 'Pathways for environmental evaluation: a walk in the (Hanging) Gardens of Babylon'. *Ecological Economics* 34, 175–193.

O'Neill, O. 1997: 'Environmental values, anthropocentrism and speciesism'. *Environmental Values* 6, 127–42.

Parker, D.J. and Handmer, J.W. 1998: 'The role of unofficial flood warning systems'. *Journal of Contingencies and Crisis Management* 6, 45–60.

Pearce, D., Markanda, A. and Barbier, E.B. 1989: *Blueprint for a Green Economy*. Earthscan, London.

Peterson, A.L. 2001: *Being Human: Ethics, Environment and Our Place in the World*. University of California Press, Berkeley, 289.

Press, M.C. and Graves, J.D. 1995: *Parasitic Plants*. Chapman and Hall, London.

Rick, C.M. and Bowman, R.I. 1961: 'Galápagos tomatoes and tortoises'. *Evolution* 15, 407–411.

Roy, A. 1999: *The Cost of Living*. Flamingo, London.

Sagoff, M. 1988: *The Economy of the Earth*. Cambridge University Press, Cambridge.

Sagoff, M. 1998: 'Aggregation and deliberation in valuing environmental public goods: a look beyond contingent pricing'. *Ecological Economics* 24, 213–230.

Smith, D.M. 2000: *Moral Geographies: Ethics in a World of Difference*. Edinburgh University Press, Edinburgh.

Snow, D.W. 1981: 'Coevolution of birds and plants'. In P.L. Forey (ed.), *The Evolving Biosphere: Chance, Change and Challenge*. British Museum and Cambridge University Press, London, 169–78.

Stoddart, D.R. 1967: 'Organism and ecosystem as geographical models'. In R.J. Chorley and P. Haggett (eds.), *Models in Geography*, Methuen, London, 511–548.

Tiffen, M., Mortimore, M. and Gichuki, F. 1994: *More People, Less Erosion: Environmental Recovery in Kenya*. John Wiley & Sons Ltd, Chichester.

Tronto, J. 1993: *Moral Boundaries: A Political Argument for an Ethic of Care*. Routledge, London.

Trudgill, S.T. and Richards, K.S. 1997: 'Environmental science and policy: generalisations and context sensitivity'. *Transactions, Institute of British Geographers* 22, 5–12.

van Andel, T.H. 1994: *New Views on an Old Planet A History of Global Change*. Cambridge University Press, Cambridge.

PART IV

FUTURES

12

'The writing's on the walls': on style, substance and selling physical geography

Heather Viles

> *It is true that in our time we have had some eminent scientific workers, who have also been masters of nervous and eloquent English. But it is not less true that the literature of science is burdened with a vast mass of slipshod, ungrammatical and clumsy writing, wherein sometimes even the meaning of the authors is left in doubt.*
>
> Geikie, 1898, in Geikie, 1905, 288

> *Science . . . is a process of intense criticism.*
>
> Dunbar, 1995, 31

12.1 Introduction

Physical geography, like all sciences, is a collective enterprise. Good communication is thus vital, both within the discipline and in representing our subject to others. Worries over our skill at communicating have intensified since Archibald Geikie's stirring address in 1898 to students of Mason University College, Birmingham, in which he laid down the importance of good writing as a key skill for budding scientists, alongside those of observation, reading, accuracy and patience. Differences in the nature of communication styles between scientists and artists are one key component of the broadening gap between the 'two cultures' identified by C.P. Snow in 1959 (although his seminal essay does not mention writing explicitly, see Snow, 1993).

In recent decades there have been extensive discussions over the place of physical geography in relation to human geography which address the reality and meaning of the 'two cultures' in a disciplinary setting (Johnston, 1986; Goudie, 1986; Worsley, 1985; and Slaymaker, 1994). Other debates affecting physical geographers also relate to issues of communication and the interfaces between different branches of science. How unified is 'physical geography' itself? Do geomorphologists and climatologists, for example, share anything more than, say, geomorphologists and historical geographers? In the United States, geomorphologists have debated their position in relation to both geography and geology (see http://main.amu.edu.pl/~sgp/gw/gggeo/gggeo.html) with some feeling that geology has 'lost interest' in geomorphology. In Britain, the reverse situation might be interpreted from the number of geology departments that are reclaiming geomorphological interests in the form of 'environmental geology'. Issues of communication are at the heart of all these discussions, and it is worth examining in more detail whether we physical geographers are making ourselves clear, both to other scientists and also to other academics, students at all levels and the wider public.

Writing and communicating about our research can be seen as the last stage of any research project, and acts to crystallize and record our ideas and findings. As such, what we write and how we write it provides a mirror of our subject as a whole. Understanding the nature of our writing, and the audiences who read it, may help us understand the changing relationship between different parts of physical and human geography and other cognate subjects. Furthermore, scrutiny of writing styles may facilitate collaboration between specialisms, as writing is our major tool of communication. In the following essay I will consider questions of what we write and for whom and discuss how our writing shapes up.

12.2 What is 'writing'?

It might be useful to examine more closely what we mean by writing. At its most basic, writing is 'a group or sequence of letters or symbols', or 'a piece of literary work done; a book, article etc.' (*Concise Oxford Dictionary*, 1990). In today's world, scientists' writing is much wider than the first definition implies, and the end products more diverse than the second definition suggests. Writing now includes images, tables, graphs, models and other things as well as simple 'text'. We now use many media as vehicles for our writing, as well as pages of paper in books and journals, such as web pages, CD-ROMs, and PowerPoint presentations. The options open to us in terms of writing on paper have also expanded hugely, and include international refereed journals, research monographs and textbooks, as well as the whole host of 'grey area' literature, which is usually taken to mean reports, conference papers and articles which are not easily obtainable. For example, research commissioned by a company into the environmental implications of a possible expansion of one of its plants would probably only be accessible by a limited number of company employees. Writing is only one of many forms of communication used in academic life and elsewhere, along with verbal presentations, images, sounds and others. Increasingly, the distinctions between forms of communication are becoming blurred with multi-media formats, for example, allowing text, data, images and lectures to be appreciated simultaneously.

Writing is not just an activity, it also creates a repository of knowledge which forms the literature of a subject. Until recently, published and printed material in book and journal format was the dominant component of the literature of any subject. However, it is now much more of a challenge to 'review the literature' on a particular topic, requiring access to online sources, as well as conventional library resources and grey area literature.

12.3 What do physical geographers write?

The phrase 'scientists must write' was used as the title of a book by Robert Barrass in 1978 and reflects a commonly held axiom that science is of no real use until it has been written or published. As Barrass puts it:

> *The literature of science, a permanent record of the communication between scientists, is also the history of science: a record of the search for truth, of observations and opinions, of hypotheses that have been ignored or have been found wanting or have withstood the test of further observation and experiment. Science is a continuing endeavour in which the end of one investigation may be the starting point for another. Scientists must write, therefore, so that their discoveries may be known to others.*
>
> Barrass, 1978, 25 (original italics)

There are several basic forms of writing produced commonly by scientists, including research reports in academic journals, books (both monographs and texts at various levels), internal reports and popular articles. Other formats include book reviews, encyclopedia entries, essays and obituaries. Technology, funding and bureaucratic structures condition what we write, encouraging some forms over others and facilitating the development of new ones. The flowering of the medium of the internet, for example, has produced a whole new genre of writing in physical geography. The home page format encourages bite-sized chunks of attractively illustrated prose and video clips, as well as downloadable data and packaged reports. The search for research funding has led to the development of several new genres – the research grant application with its gritty prose crammed into proscribed space; the research report with its one-page executive summary for busy funders to flick over, and page after page of partly digested results to illustrate the progress (and need for further funding!); and the glossy research project flier – complete with acronyms and logos galore. Within the context of academic physical geography in Britain, the recent Research Assessment Exercise (RAE) has encouraged certain types of writing over others, with many departments instructing their physical geographers to write in only one genre, i.e. the research paper in an international refereed journal. Other types of writing have been regarded as inferior for RAE purposes.

12.4 For whom do physical geographers write?

There is an audience for every piece of writing even if, as in the case of many PhD theses, this audience may be very small. I'm sure that no one apart from my PhD examiners, a few misguided students and my mother have ever read my thesis. It probably only made a real impression on my mother, who couldn't understand why I made three years of enjoyment out of something so impenetrable. Physical geographers have become very adept at targeting their writing at a specific audience, as in the case of grant applications, where buzz-words such as 'timeliness', 'novelty' and 'inter-disciplinary' pepper even the dullest and most obscure proposal. Each of the genres of physical geography writing I have discussed in the preceding section have a particular audience in mind. The refereed journal article is written to impress one's peers, the textbook chapter is written to explain a subject to students, and consultancy reports are written to satisfy clients that work has been done and questions answered. As J.M. Smith (1996) indicates, the audience is important in conditioning the style and substance of what we write. As he puts it, 'To satisfy an audience, and earn its trust, the writer must confirm their prejudices and respect their preferences' (Smith, 1996, 20). Are we targeting the right styles of writing and the right audiences? Or have physical geographers lost the plot?

12.5 Have we got our writing right?

There has been much discussion of the style and substance of writing in human geography (Billinge, 1983 and Smith, 1996), but physical geographers have expressed less of an interest in what we write and how and for whom we write it. Smith (1996) illustrates the many different types of writing style produced by geographers and shows how the preference of the reader influences how they assimilate the material presented. For example, geographers who write using the trope (or figure of speech) of irony are likely only to be appreciated by those readers who think in those terms. Often, physical geographers (and probably some human geographers as well) will read one of the more esoteric pieces of human geography writing and dismiss it as gobbledygook, largely because of the style of writing, which seems to baffle with long words and slippery concepts. Clearly we are not the intended audience. In comparison, many human geographers become depressed by the dry, factual presentation and lack of discussion of ideology in much physical geography writing. Again, this implies that they are not the target audience. As Dunbar (1995) indicates, such differences in language are also found within different branches of science, with some disciplines using words and concepts in wholly different ways to other scientists. In the following paragraphs I will examine first the issues of writing for what we currently see as the target audiences of our prose and, second, larger questions of whether we are writing to the right audiences.

To many scientists, style is subordinate to substance. If physical geographers consider writing style, they often adopt the position of Wright (1947) who claimed that 'the style of scientific exposition should be clear, simple and concise'. Barrass (1978, 32) exhorts us that 'scientists should write direct, straightforward prose, free from jargon, verbosity and other distracting elaborations'. He goes on to recognize eight key elements of good scientific prose, i.e. explanation, clarity, completeness, impartiality, order, accuracy, objectivity and simplicity. These elements echo the norms of scientific behaviour proposed by R.K. Merton in 1942 (communalism, universalism, disinterestedness, originality and scepticism, see Ziman, 1984, for a useful review). Most physical geographers adopt such guidelines, and thus aim for a bland, economical style, structuring papers and reports under headings of aims, methods, results, discussion and conclusions.

However, we should examine whether we have got it right. To what extent does writing in physical geography conform to these rigorous standards?

Although physical geographers do not engage in the flowery rhetoric beloved of many human geographers, we are often guilty of using a plethora of technical terms without adequate definition or explanation. We also fail to accommodate the 'rampant math anxiety' as Phillips (1999) puts it, mixing textual and numerical descriptions in a way which can alienate the uninitiated. Frustratingly, many papers fail to give adequate information on the exact methods used and data collected, making reproduction of the study impossible. Furthermore, many papers are rather selective in their presentation of data, conclusions and reference to previous work. In the current, competitive climate of research, with ever more pressure on individuals to publish, is physical geography writing becoming better or worse? On the one hand, increasingly tough peer review procedures are no doubt honing our phraseology and removing our idiosyncrasies. On the other hand, the pressure to publish seems to be producing more and more specialized papers whose target audience is becoming ever more narrowly focused. Thus, most fluvial geomorphology papers are now almost incomprehensible to geomorphologists with other interests.

Looking back at things written by physical geographers which have made an impression upon me, or which have provoked me in some way, it is clear that style is as important as substance. I've picked two papers to illustrate the point; one which came out whilst I was an undergraduate and one which was published whilst I was doing post-doctoral research and wondering (not for the last time) whether an academic career in geography was for me. In 1979, Stanley Schumm published a paper in *Transactions of the Institute of British Geographers*, entitled 'Geomorphic thresholds: the concept and its applications'. The paper immediately follows what has become one of the classic papers in geomorphology, that by Denys Brunsden and John Thornes on 'Landscape sensitivity and change'. As an undergraduate I remember finding both difficult and challenging, but having a clear preference for Schumm's paper which simply made more sense. The message that Schumm conveyed was important and the examples he gave pertinent and easy to follow. In 1987, David Stoddart published a paper, also in the *Transactions of the Institute of British Geographers*' which truly inspired me, entitled 'To claim the high ground: geography for the end of the century'. What a rallying cry! The text of a lecture given at the University of California at Berkeley, the paper reads as though David Stoddart is standing there speaking to the reader. He is thoughtful and scholarly, but at the same time opinionated and provocative. Looking through the well-thumbed copy of the journal from my departmental library, I note that he still continues to provoke and surprise. In the margin next to Stoddart's sentence in which he reveals some of the things that he has read which have inspired him, 'I have myself a special affection for Cressey's *China's Geographic Foundations*, for Trewartha's *Japan*, for Spate's *India and Pakistan*' (Stoddart, 1987, 331) someone has written 'You're mad'. This paper provides a useful basis for tutorial discussions as it always produces some strong reactions, but it continues to bring a smile to my face each time I read it.

It is difficult to quantify changes in writing, and it is even more difficult to quantify the key aspects of writing which make it attractive to certain audiences. However, some comparisons between papers published in *Earth Surface Processes and Landforms* in 1976, 1986 and 1996 will serve as a rough guide to trends in physical geography papers (see Figure 12.1). The number of papers in the journal has increased from 28 a year in 1976, to 72 in 1996 (excluding those in the *Technical and Software Bulletin* which came out as issue no 11). Over this period, the length of the titles has increased (with the longest in both 1976 and 1986, for what it's worth being written by Steve Trudgill and co-workers!). Length isn't everything, clearly, but it is also instructive to see that the nature of titles has also changed: from bold assertions such as 'A scree slope rockfall model' or 'The morphology of riffle-pool sequences', to more timid and less snappy versions such as 'On the necessity of applying a rotation to instantaneous velocity measurements in river flows', or the colon-cursed 'Stone cover on desert hillslopes: extent of bias in diameters estimated from grid samples and procedures for bias correction'. At the same time, the number of authors per paper has increased from a mean of 1.4 in 1976 to 2.3 in 1996, suggesting, at least in theory, that papers are becoming increasingly written by teams, with individual style becoming submerged. A measure of the complexity of papers might be given by the number of pages, figures and tables involved – all of which have shown an increase over the period. More detailed analysis of both easily quantifiable and less easily quantifiable aspects of the nature of physical geographers' writings is needed to provide a better description of the trends. However, the general point remains that, at a time when accessibility is a key social issue, and when scientists are increasingly becoming aware of the interconnectedness of ideas, it would seem a good idea for us to re-evaluate our writing to ensure that we do not reduce our audience even further.

Are there other audiences for whom we are not writing? Or for whom we are not adapting our style sufficiently to make our arguments persuasive? The pressure of the

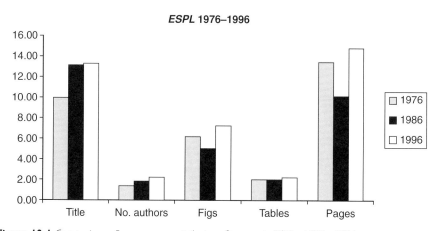

Figure 12.1 *Comparison of some mean attributes of papers in* ESPL, *1976–1996.*

RAE in Britain has reinforced a decline in the amount of writing by physical geographers for A level audiences and for wider readership. Even textbooks for university-level students are frowned upon by many as being 'a waste of time' and 'no use for the RAE'. Similarly, reflective articles on ideas and practice of physical geography and review articles (which can provide an invaluable introduction to a new field) find few outlets. Thus we are tending to write more and more detailed research material (from grant proposals, through internal reports to final refereed journal articles) and less and less of any other genres. As a community of physical geographers we should consider what sort of literature we want to develop for our needs and those of future geographers – are research papers enough? Are we giving enough attention to promoting the best of other genres? Is our subject sustainable without a broader array of writing and types of communication? How much this matters depends upon one's viewpoint of our individual commitments to the collective enterprise of science, and the need for us to communicate our science to non-scientists and the scientists of tomorrow.

Writing is both a symptom and a cause of 'the trouble with science'. How we write and for whom we write conditions people's view of us and our science. As physical geographers we should reflect more on our writing and continue to improve the quality and diversity of our outputs. We should be just as critical of how we write up our science as of how we do our science.

References

Barrass, R. 1978: *Scientists Must Write: A Guide to Better Writing for Scientists, Engineers and Students.* Chapman and Hall, London.

Billinge, M. 1983: 'The Mandarin dialect: an essay on style in contemporary geographical writing'. *Transactions, Institute of British Geographers* 8, 400–20.

Dunbar, R. 1995: *The Trouble with Science.* Faber and Faber, London.

Geikie, A. 1905: *Landscape in History, and Other Essays.* Macmillan, London.

Goudie, A.S. 1986: 'The integration of human and physical geography'. *Transactions, Institute of British Geographers* 11, 454–458.

Johnston, R. 1986: 'Four fixations and the quest for unity in geography'. *Transactions, Institute of British Geographers* 11, 449–453.

Phillips, J.D. 1999: *Earth Surface Systems.* Blackwell, Oxford.

Schumm, S.A. 1979: 'Geomorphic thresholds: the concept and its applications'. *Transactions, Institute of British Geographers* 4, 485–515.

Slaymaker, O. 1994: 'Is geography sustainable without geomorphology?' *The Canadian Geographer* 38, 291–300.

Smith, J.M. 1996: 'Geographical rhetoric: modes and tropes of appeal'. *Annals, Association of American Geographers* 86, 1–20.

Snow, C.P. 1993: *The Two Cultures*, with introduction by Stefano Collini. Cambridge University Press, Cambridge.

Stoddart, D.R. 1987: 'To claim the high ground: geography for the end of the century'. *Transactions, Institute of British Geographers* 12, 327–336.

Worsley, P. 1985: 'Physical geography and the natural environmental sciences'. In R.J. Johnston (ed.), *The Future of Geography.* Methuen, London, 27–42.

Wright, J.K. 1947: 'Terrae incognitae: the place of the imagination in geography'. *Annals, Association of American Geographers* 37, 1–15.

Ziman, J. 1984: *An Introduction to Science Studies. The Philosophical and Social Aspects of Science and Technology.* Cambridge University Press, Cambridge.

13
Conclusion: contemporary meanings in physical geography

Stephen Trudgill

Peter Sims concludes his contribution to this book on an upbeat note concerning 'the vitality of the subject matter and in the willingness displayed by researchers to explore new initiatives as well as the variations on more traditional themes.' So perhaps I am being unduly pessimistic in feeling that convergent thinking is what we have to avoid (Chapter 2). Sims continues, 'Long may this continue in either the reductionist or holistic approaches to geomorphology, although I have to admit to a preference for the latter', which is a theme that Mike Urban and Bruce Rhoads take up in their later chapter, as does Keith Richards.

The theme of the role of autobiography is taken up by Tim Burt and Tim Bayliss-Smith. Tim Burt records clearly his formative early years and the importance of an enthusiastic and able teacher in the inspiration to study physical geography. He feels that now risk and uncertainty may well become future themes in physical geography, but also reaffirms the basic, and original, inspiration that comes from the landscape itself. Tim Bayliss-Smith portrays graphically how an idea can seize the research worker, so much so that he attempted to quantify almost every flow in an island ecosystem! He concludes by warning of the dangers of 'persuasive rhetorics' and the need for recognizing their limitations as well as their strengths.

In many ways, Chris Keylock picks up the baton from Tim Bayliss-Smith and runs with it by noting the strength of paradigms and the inertia of stereotypes. He prefers to think of a fluid hierarchy of scientific revolutions. His chapter made me think of the 'drowning embrace' written of by Laurens van der Post:

> *History teaches that men as a rule do not break with specialised processes which apparently served them well, without the help of disaster . . . [which] . . . free(s) the human spirit from the drowning embrace of exclusive conditions.*
> *Laurens van der Post, 1978, Jung and the Story of Our Time, Penguin.*

Once again, I began to muse on the relationship between free thinking and paradigms, seeing paradigms as a 'drowning embrace', and even wondering whether there ever is such a thing as free thinking. Perhaps, however, there is a tendency to become too locked in the paradigm of paradigms and maybe they are more illusory than might be apparent. However, and refreshingly, Roy and Lane conclude that new hypotheses continue to be tested experimentally in a cross-breeding of approach involving interactions and exchanges of ideas with the recognition of their individual as well as collective benefits. Further, David Favis-Mortlock and Dirk de Boer stress the importance of unifying principles and universal applicability.

John Thornes, whose book, *John Constables Skies – A Fusion of Art and Science*, I have greatly enjoyed, grasps the issue of science and society and the authors again write refreshingly of the cultural significance of climatology as one aspect of physical geography. Gerardo Bocco indeed leads us into a discussion of 'landscape wisdom', where the societal significance of physical environmental understanding is very great. Finally, Mike Urban and Bruce Rhoads write of an integrated geography, and especially about the role of cultural constructs of nature, and then Keith Richards extends the theme by thinking about an applicable human ecology where concerns with humanity and the environment are appropriately balanced and which is, at the same time, a form of moral geography.

In this, the concluding section, Heather Viles focuses on the topic of communication, feeling that improving not just the quality but, more importantly, the diversity of our writings is important in order to engage a wider audience.

Overall, it is clear that while the endeavour of the scientific testing of our ideas about how the natural environment works and the search for pattern remain central, to my mind three other themes emerge in the contemporary meanings of physical geography.

1. The importance of the actual physical landscape itself – as an inspiration, a sense of awe and wonder.

2. A sense of the importance of a wider engagement not only with human geographers and applied environmental science, but also, and perhaps more importantly, an engagement with society as a whole, and especially the cultural significance of physical geography with the understanding of the natural environment which it has to offer.

3. A sense of caution about how our endeavours in physical geography evolve and, in particular, the importance of being inspired by teaching while avoiding the 'drowning exclusivity' of existing ideas.

The first is admirably addressed by Bate (2000) in *The Song of the Earth* and the concept of ecopoetics. He writes about watching a skylark rising: 'My heart leaps up. But my mind has fallen into knowledge: a biologist will be able to explain to me why the lark rises (but) the freedom of the lark is … in my imagination'. He writes further about 'the richest thoughts and feelings' and 'of reunifying the world' of feelings through metaphor. Thus, in physical geography, we may both write a scientific treatise on glaciers but also appreciate, and in many ways be motivated by, metaphor in the landscape. W.H. Auden's poem 'In Praise of Limestone' talks of:

> *.... rounded slopes*
> *With their surface fragrance of thyme and, beneath,*
> *A secret system of conduits: hear the springs*
> *That spurt out everywhere and chuckle...*

and

> *when I try to imagine a faultless love*
> *Or the life to come, what I hear is the murmur*
> *Of underground streams, what I see is a limestone*
> *landscape.*

More directly geomorphological, there is:

> *Crawl on old ice worm, from the solemn hills;*
> *Press deep thy burrowing snout among the stones;*
> *Mutter and murmur with thy turbid rills,*
> *And crush the old earth's bones ...*
> *Gnaw, grind the patient cliffs with ravenous teeth ...*
> *Haste thee, for thou art destined to decay ...*
> *And though shalt not prevail.*
>
> A.C. Benson 'By the Glacier' (Newsome, D., 1980)

It is clear that the landscape has human meaning through metaphor, and this leads us on to the cultural significance of the physical environment covered by the second point above. This is a growing awareness, as John Thornes suggests in Chapter 8 in this book.

The third point about how our ideas evolve is one I would like to return to in conjunction with the interaction between personal and research meanings, and, in one of the spirits of this book, autobiographically, as follows.

When I was an undergraduate at Bristol in the late 1960s, I remember having two lectures on erosion surfaces – a subject which we came to dub 'geomythology'. With hindsight, this seems like rather a cruel judgement since there was, in essence, an elegant intellectual argument about how landforms must have evolved given the logic of sequences and available evidence. But this judgement, I am sure, stemmed from our glee on becoming postgraduates and actually being able to measure things like erosion rates which would surely test the hypotheses of landform evolution which we had learnt about earlier.

The ethos was a youthful one of being able to challenge the established views with 'real' measurements rather than supposition. In those heady days, surrounded by David Harvey, Peter Haggett, Andy Cliff and Mike Kirkby, and when conventionally I had hair down over my shoulders and joined CND, I remember Dingle Smith's tale of how a chemist, David Mead, had met with Dingle, Marjorie Sweeting and Professor Tratman through the Bristol Speleological Society and enlightened them about EDTA titrations through which you could actually measure the amount of calcium carbonate in solution in runoff waters (Smith & Mead, 1962) – and even use the data to calculate erosion rates (Atkinson & Smith, 1976). The micro-erosion meter was invented (High & Hanna, 1970) and we really went to conferences simply to announce results about erosion rates. Soon I became adept at cutting up bits of rock, placing them in the environment and re-weighing them in order to study erosion (Trudgill, 1975) following Newson's (1970, 1971) work. It was *de rigueur* to do water traces and the laboratory bubbled with cauldrons of *Lycopodium* spores and the fluorometer hummed into the night as we traced karst waters.

I remember the mantra: 'if it moves, measure it', and pretty soon we were measuring everything from bedload and suspended sediment to the rate of bioerosion by boring molluscs and the infiltration of water into soils (Trudgill, 1983). As I look back, it seems that the original questions, which were about how fast landforms actually eroded – and therefore whether the hypotheses about their evolution were actually feasible in the time available, were almost completely forgotten. Measurement and the study of process (Embleton & Thornes, 1979) almost seemed enough themselves. Heather Viles (1992, 187–8) graphically captures the joys of process measurement on a salt marsh, using filter papers to measure sedimentation rates, aided by her mother holding the surveying staff. There then seemed a natural progression from measurement to quantification within a systems context (Chorley and Kennedy, 1971) and as applied to soil and vegetation (Trudgill, 1977).

Before long, one had to get serious about grants and applicability so the techniques became, for me, applied to tracing soil water and the measurement of nitrate leaching, which seemed more fundable and relevant than the study of the evolution of landforms. Interestingly, work on the losses from nitrate from agricultural land brought me in to contact with farmers and I rapidly learnt about the importance of perceptions and the human dimensions of environmental management. This is so much so that I now lecture at Cambridge on attitudes and values (Trudgill, 2001) as much as on environmental biophysical processes.

What does this brief autobiographical sketch tell me? First, we read of paradigms and paradigm shifts – it seems that these are very real. Why do I not go to a conference and give a paper on the geomorphological evolution of the South Downs in 2002? The answer is that it is not an 'in' topic and far from fundable. Why did I think that standing in the wet taking measurements was a worthwhile pursuit in the 1960s and 70s? The answer was that that was what everyone else was doing and that I was engaged in the endeavour of the times. Chapter 4 in this book, by Tim Bayliss-Smith, rehearses as much when, inspired by David Stoddart, he spent time measuring everything he could about an island ecosystem. The sociology of academia is very strong. Why, then did I later go in for nitrate and other environmental work? The answer is that fundability became even stronger than the peer pressure from the academics and researchers around me. That much should be obvious. The questions arise, however, of how far we should be sensitive to such pressures. Should they be resisted or are they inevitable? As a postgraduate, I never asked such questions; process measurements seemed a natural thing to do – it equated in many ways with sociability. But now one worries about fundability becoming a constraint on the intellectual endeavour of enquiry.

Personally, I was predisposed to things biological and fieldwork, having spent my earlier youth outdoors birdwatching and roaming round Norfolk with the Norfolk and Norwich Naturalists Society; so in many ways it was natural for me to engage in fieldwork and, later on, in environmental issues, and I guess it was a putative recognition of kindred spirits which made me engage with the process fieldwork. My greatest spur was a field trip, at the end of my first year, to the Burren in Co. Clare, where I spent six weeks with postgraduates happily helping with their speleological fieldwork. I was deriving a private, personal meaning from being outdoors surrounded by plants, wildlife and scenery, writing nature poems, which I showed to few people, and enjoying the physical experience and companionships of caving. I was also deriving a more public, research meaning from being part of something greater than me – an academic endeavour which was putatively at the forefront of part of the quantitative revolution of measurement and process study.

For my PhD on limestone erosion on the remote coral atoll of Aldabra in the Indian Ocean, I know that I was working in the contexts set for me by the 'process measurement' ethos of Bristol. In an interdisciplinary team of biologists and geologists, I was constantly thinking of supervisors (Dingle Smith, David Stoddart), other postgraduates and British geomorphologists with their zeal for measurement and finding out process–form relationships, bringing this, my context, to the team while also learning from the others about geology and biology. This was nested in a wider context of coral atolls, discourses on islands, exploration and discovery, so I was often reading and referring to earlier expeditions and Darwin in my mind. Even wider, and also more fundamentally, I was nested in a personal love for nature and wildlife, especially in terms of the indigenous flora and fauna, like the Giant Tortoise and flightless Rail. My Aldabra diaries are full of notes and field sketches of landforms with deductions about probable formation, but also there are loads of scribbled poetic writings which give a flavour of my personal experience:

A quiet moment of unsurpassed peace, sitting on the beach, amongst the Casuarina pines, a small fire going, glowing red in the dark, throwing a pale glimmer over the sand. Looking out over the still pool of the sea in which a bright star shone a reflection. Listening to the surge of the waves on the reef in the dark beyond and feeling the warm breath of wind under the vast sky and millions of stars.

And then:

> There is much that man does love
> that does not much love man.
> The vast blue inhumanity
> of boundless sea and sky.
> Vast inhumanity, of elements of sky, of sea
> makes my light
> my spark of life shine brighter.
>
> Trudgill, 2001, 106

My personal motivations are clear – a deep, poetic personal enjoyment of wild nature and the wider, higher context of science and finding out, clearly placed in thoughts of expeditions, Darwin, islands and so on, and the specific location of myself in the process geomorphology ethos of Bristol at the time.

It is the dialogue between these two sets of meanings – private/personal and public/research – which fascinates me. Clearly the two did not conflict for me in the early days, and I can be reasonably clear about the motivations for undertaking particular kinds of research.

It would seem also that I need not feel too bleak about the pressures of fundability in that I have always been able to find personal meanings, even when evolving through to applied nitrate work which was more driven by the prospects of external funding than internal volition. The fact is that it is essential and unavoidable to find personal meanings, however strong the external pressures are, even if we might disagree with them. Thus I found the applied nitrate work did have an inner meaning in terms of environmental issues and that the personal aspects of meaning, involving 'being outdoors', the teamwork and the 'ecological discourse', are there all the way through – though increasingly, perhaps, research meaning becomes separate from personal meaning, as research is driven by outside pressures, rather than, and perhaps regrettably, inner thought, and motivated by personal inspiration. Here the personal meanings might thus be given rather than internally generated, but even when given other personal meanings can be retained intact. Thus the dialogue between personal meanings and public meanings can always be rationalized, but where does this leave the all-important flow of new ideas?

Was it right that 'we all did erosion surfaces' and then 'we all did process measurements' and then 'we all found applied, fundable topics'? Would it not have been intellectually healthier if there had been a legitimized plurality of approaches? Or does each generation need to pour scorn on previous ideas as we cruelly did to 'geomythology'? It certainly seems healthy to question previously accepted ideas and this can usefully be in a way which is generational. I guess I am slightly worried both about the ways subjects swing around in paradigm shifts in rather a 'herd instinct' manner, and also about the way fundability means that new knowledge is derived from avenues which are directed by current conventional wisdom. I can only conclude that we, and the subject we profess, are well served by mavericks and debate rather than by consensus. But still there is the social imperative of 'belonging'. As I look back, this sense was very much a strong motivation, so perhaps a plea for new ideas does involve 'the way the subject is going' as much as where individuals might take us. Even for the maverick, the need for 'approval' is very strong.

This is not to deny the importance of scrutiny but to emphasize the point about not stifling originality. Haggett (1965) uses two quotations which are relevant here: 'That there is more order in the world than appears at first sight is not discovered till order is looked for', from Sigwart (Haggett, 1965, 2). Haggett continues further: 'The "seeing eye" beloved of the late S.W. Wooldridge, is a necessary part of our scientific equipment in that pattern and order exist in knowing what to look for, and how to look.' Haggett also quotes from Popper: 'Bold ideas, unjustified anticipations, and speculative thoughts, are our only means of interpreting nature ... Those among us who are unwilling to expose their ideas to the hazard of refutation do not take part in the scientific game' (Haggett, 1965, 29). I happily endorse the last point, but emphasize that when we are looking for order, and using the 'seeing eye', this can both enrich our experience and constrain it if we let one interpretation preclude others. Most fundamentally, *we do need the 'bold ideas, unjustified anticipations, and speculative thoughts' to advance the subject*, and we should never used scrutiny to stifle these or let current ideas constrain our seeing. Approval may only mean that some interpretation fits with current thinking and given meanings, which is a recipe for stifling new directions.

So, to re-visit the points I made above, physical geography seems to me to include at least three discourses in terms of contemporary meanings:

1. **Understanding process and forms** – how things work, the search for pattern, logic, sequences and relationships. Here we should be open-minded as much as finding confirmation for pre-recognized patterns.

2. **Applicability and relevance.** This involves cultural attitudes to nature – from environmental management to working with nature and letting nature take its course, as well as the aesthetic experience. What is important is the way our knowledge is seen as significant and can be used, interpreted and assessed by different people with a range of terms of reference in a number of different contexts. In the broadest possible way, from scientific interpretation to poetic metaphor, physical geography is socially and culturally significant.

3. **Motivations** – the personal and research meanings derived from the research. Why do we research on what we do? How have we been influenced by others? It seems there is a mixture of personal interests, our own autobiographies and the intellectual contexts which others have framed for us but which we eventually make our own.

The first point is, perhaps, well rehearsed but it seems that the second is a growing area of interest whatever way it is expressed, such as in the contexts of *integration, human ecology, applicability, social relevance* or *cultural significance*. The third must be worthy of further attention. In a paper on 'Gender, culture and astrophysical fieldwork' by Pang (1996), which examines the contextual situation of fieldwork, the author writes: 'theorising, experimenting and instrument building are the central activities of science' but 'we should broaden our analytical perspectives to include accounts of what it feels like to build a camp, handle a new instrument . . . , deal with solitude and worry in the field, master a difficult technique, make a discovery, accept a failure.' By considering **'living and feeling people** (doing the research)**, we can discover links between social life, culture and work that are otherwise edited and unrecorded, and build a richer and more complete, and even more passionate, picture of scientific practice.'**

THE GLACIER

The mountains have a peace which none disturb;
The stars and clouds a course which none restrain;
The wild-sea waves rejoice without a curb,
And rest without a passion; but the chain
Of death, upon this ghastly cliff and chasm.
Is broken evermore, to bind again,
Nor lulls nor looses. Hark! A voice of pain
Suddenly silenced; a quick-passing spasm.
That startles rest, but grants not liberty –
A shudder, or a struggle, or a cry –
And then sepulchral stillness. Look on us,
God! Who hast given these hills their place of pride,
If death's captivity be sleepless thus,
For those who sink to it unsanctified.

John Ruskin

This quotation, for me, brings together many aspects of the scientific endeavour, academic contexts, cultural significance and personal meanings which are interwoven in the contemporary meanings of physical geography. We may know what it is to undertake a scientific study whereby we can understand a glacier in terms of origin, physical operation, hydrological processes and geomorphological and environmental significances, but we may also have that richness of being where we can articulate the whole experience of what it is like to both understand and stand on a glacier.

References

Atkinson, T.C. and Smith, D.I. 1976: 'The erosion of limestones'. Ch. 5 in T.D. Ford and C.H. Cullingford (eds.), *The Science of Speleology*. Academic Press, London.

Bate, J. 2000: *The Song of the Earth*. Picador, London.

Chorley, R.J. and Kennedy, B.A. 1971: *Physical Geography. A Systems Approach*. Prentice Hall, London.

Embleton, C. and Thornes, J. (eds.), 1979: *Processes in Geomorphology*. Arnold, London.

Hagget, P. 1965: *Locational Analysis in Human Geography*. Arnold, London.

High, C.J. and Hanna, K.K. 1970: *A Method for the Direct Measurement of Erosion on Rock Surfaces*. British Geomorphological Research Group, Technical Bulletin, 5.

Newsome, D. 1980: *On the Edge of Paradise: A.C. Benson the Diarist*. Murray, London.

Newson, M.D. 1970: 'Studies n the chemical and mechanical erosion by streams in limestone terrains'. Unpublished PhD thesis, University of Bristol.

Newson, M.D. 1971: 'The role of abrasion in cavern development'. *Transactions of the Cave Research Group of Great Britain*, 13, 101–7.

Pang, A. S-K., 1996: 'Gender, culture and astrophysical fieldwork'. In H. Kuklick and R.E. Kohler (eds.), *Science in the Field*. Chicago University Press, Chicago.

Smith, D.I. and Mead, D.G. 1962: 'The solution of limestone with special reference to Mendip'. *Proceedings of the University of Bristol Speleological Society*, 9, 188–211.

Thornes, J.E. 1999: *John Constables Skies – A Fusion of Art and Science*. University of Birmingham Press, Birmingham.

Trudgill, S.T., 1975: 'Measurement of erosional weight-loss of rock tables'. *British Geomorphological Research Group, Technical Bulletin*, 17, 13–19.

Trudgill, S.T. 1977: *Soil and Vegetation Systems*. Oxford University Press, Oxford.

Trudgill, S.T. 1983: *Weathering and Erosion*. Butterworth, London.

Trudgill, S.T. 2001: *The Terrestrial Biosphere: Environmental Change, Ecosystem Science, Attitudes and Values*. Pearson, London.

Viles, H.A. 1992: 'Physical geography fieldwork'. Ch. III.9 in A. Rogers, H. Viles and A. Goudie (eds.), *The Student's Companion to Geography*. Blackwell, Oxford. 187–195.

Index